Soil Microbiome of the Cold Habitats

This book focuses on cold habitat microbes as a potential source of elite enzymes and secondary metabolites to meet the growing demands of the pharmaceutical, food and biotechnological industries. Microbes living in such extremely cold conditions are reported to produce various biomolecules with potential biotechnological applications.

The book overviews recent research trends to discover such important biomolecules and also suggests future research directions to discover such elite novel biomolecules.

Salient features:

- Covers studies on various biotic communities and abiotic components of the soil of terrestrial habitats, with a focus on cold habitats
- Discusses various 'Omic' approaches: metagenomics and meta-transcriptomics
- Lists adaptation strategies adopted by cold-adapted microbes
- Highlights various biotechnological and industrially important biomolecules produced by cold-adapted microbes
- Explores the role of microbial biofilm in the degradation of microplastics in cold habitats

Soil Microbiome of the Cold Habitats

Trends and Applications

Edited by
Puja Gupta
and Mohd. Shahnawaz

CRC Press
Taylor & Francis Group
Boca Raton London New York

CRC Press is an imprint of the
Taylor & Francis Group, an informa business

Designed cover image: © Shutterstock

First edition published 2024
by CRC Press
6000 Broken Sound Parkway NW, Suite 300, Boca Raton, FL 33487-2742

and by CRC Press
4 Park Square, Milton Park, Abingdon, Oxon, OX14 4RN

CRC Press is an imprint of Taylor & Francis Group, LLC

ISBN: 9781032406305 (hbk)
ISBN: 9781032406336 (pbk)
ISBN: 9781003354031 (ebk)

DOI: 10.1201/9781003354031

Typeset in Times
by Deanta Global Publishing Services, Chennai, India

Each editor dedicates this book to their respective parents, better halves and kids for their kind support and trust

Contents

About the editors

Dr. Puja Gupta is working as an Assistant Professor of Biotechnology at the Regional Institute of Management and Technology, Mandi Gobindgarh, Sirhind, Punjab, India and is the recipient of various fellowships: DST-Women Scientist-A (funded by Department of Science and Technology, Govt. of India), CSIR-RA (funded by Council of Scientific and Industrial Research (CSIR), Govt. of India), CSIR-Senior Research Fellowship (Funded by Council of Scientific and Industrial Research (CSIR), Govt. of India). Besides these prestigious fellowships, she has worked as a Senior Research Fellow in Department of Biotechnology- and National Innovations on Climate Resilient Agriculture-funded projects at the University of Jammu and Sher-e-Kashmir University of Agricultural Sciences and Technology of Jammu, respectively. She qualified National Eligibility Test for Lectureship in Life Sciences under the Council of Scientific and Industrial Research, New Delhi, India, Government of India in December 2007. She has 5 years of academic experience and more than 10 years of research experience. She was awarded a women scientist award by the Society for Scientific Development in Agriculture and Technology, Meerut, in 2020, and the Young Researcher Award by the Institute of Scholars in 2021. She has published ten research/review articles as a first author in journals of repute. She has published eight book chapters and three books with publishers of international repute. Her research interests include metagenomics, microbiology, microbial genetics and plant–microbe interactions.

Dr. Mohd. Shahnawaz is currently working as an Assistant Professor on a contract basis at the Department of Botany, University of Ladakh, Kargil Campus, Khumbathang-194105, Kargil, Ladakh UT, India. Previously, he was selected as an Associate Professor at Biofuels Institute, School of the Environment and Safety Engineering, Jiangsu University, Zhenjiang, China (2021) and as a Post-Doctoral Fellow at Yeungnam University, Gyeongsan, North Gyeongsan, South Korea (2020). He is the recipient of various fellowships: National Post-Doctoral Fellowship, funded by the Department of Science and Technology – Science and Engineering Board, Government of India; University Grants Commission-Maulana Azad National Fellowship for Minority Students Junior Research Fellowship and Senior Research Fellowship, funded by the University Grants Commission and the Ministry of Minority Affairs, Government of India; and the University Grants Commission Basic Scientific Research Fellowship (previously known as the University Grants Commission Meritorious Fellowship), funded by University Grants Commission, Government of India. He has several years of teaching and post-doctoral experience. In 2019, he was awarded Young Scientist Award by the VD Good Professional Association at Visakhapatnam, India. He has served as a reviewer, editor and guest editor for several national and international journals. He has published more than 22 research articles/reviews, 11 book chapters and 8 books with publishers of international repute. His research interests are in ecology, microbiology, bioremediation, plant secondary metabolites, integrative medicine and drug discovery.

Preface

Microbes are present everywhere, from extreme cold to hot environments. Depending upon the type of environmental stress resisted by microbes, these are grouped according to temperature (high temperature: thermophilic, low temperature: psychrophiles), pH (acidic: acidophiles), salinity (high salinity: halophiles), etc. Among the various harsh habitats mentioned here, the microbes surviving at low temperatures, i.e., cold habitats, have attracted special attention from researchers for study. Cold habitats cover the majority of the Earth. Polar habitats like Antarctica and the Arctic, the Tibetan region and the Great Himalayas are among the cold habitats. These extreme environments are not very appropriate for the survival of human beings. Despite such extreme cold conditions, microbes thrive and occupy every corner. It is quite surprising how these microbes survive in such extreme environmental conditions. It has been documented in the literature that most microbes are the same everywhere; it is only a fraction of microbes that are endemic to a specific environment. The literature survey reveals that microbes have developed some adaptational features to withstand harsh environmental conditions. The adaptations are at the structural and functional levels. At the structural level, it could be a change in the percentage and abundance of certain amino acids. At the functional level, it could be the production of specific enzymes, secondary metabolites or other biomolecules. Such biomolecules can be exploited for various pharmaceutical and biotechnological industries.

Microbial diversity is a huge topic, and recent high-throughput sequencing techniques have played a vast role in generating data and exploring this hidden treasure. There is a need to dispense the data generated by the researchers to other scientists under the following lines: (i) to cover studies on various biotic communities and abiotic components of the soil of the terrestrial habitats, with much focus on cold habitats; (ii) to highlight various methods/techniques employed to explore microbial diversity, as well as the various biogeochemical pathways that regulate the soil microbial processes; (iii) to broaden the scope of the soil microbiome of cold habitats with various approaches already reported in the literature, i.e., metagenomics, metatranscriptomics and meta-proteomics, to get a better idea about active communities.

So, various subject experts at the international level were contacted, and the most important chapters were screened and accepted based on peer-reviewed reports. Accordingly, only 16 chapters were finalized to cover the major theme of the book. The chapter-wise synopsis of the book is as follows.

The opening chapter (Chapter 1) of the book overviews the physicochemical and biological properties of soil in cold habitats and the adoption of different strategies for maintaining soil health. Chapter 2 mentions soil as a host to various biotic communities. Chapter 3 discusses the effect of soil characteristics on soil microbial biomass and diversity. Chapter 4 highlights mainly the microbial diversity of five major cold deserts of the world, namely, the Gobi Desert; the

Taklamakan of Eurasia; the Namibian Desert in Africa; the Drass Desert of India; and the Atacama Desert of South America. Chapter 5 discusses the diversity of potassium and iron solubilising bacteria in trans-Himalayan soil. Chapter 6 discusses the potential and responses of microfungi to the dynamics of cold environments. Chapter 7 mentions the microbial diversity of rhizosphere soils of the Indian Himalayan regions. Chapter 8 mentions the soil microbiome responsible for enhanced crop productivity. Chapter 9 reports the impact of microbial biofilm community shifts on the degradation of microplastics in cold habitats. Chapter 10 portrays varied adaptive strategies employed by psychrophilic bacteria to live in cold habitats alongside their diversity in various spheres of planet Earth. Chapter 11 pools the studies on the production of secondary metabolites by some of the important psychrophilic soil microbes to report the important drug molecules. Chapter 12 describes the application of 'Omics' technologies to study soil microbiomes. Chapter 13 describes the ecological role of psychrophiles and the mechanism of adaptation in various cold ecozones. Chapter 14 describes the ecology and remediation of soil as a step toward modern soil biotechnology. Chapter 15 presents a new approach to explore the psychrophilic soil microbiome through machine learning. Chapter 16 mentions mass spectrometry as an emerging analytical technique for agriculture in a changing climate.

Puja Gupta
Punjab, India

Mohd. Shahnawaz (Khakii)
Kargil, India

Acknowledgements

We bow our heads to Almighty God for giving us the courage and patience to complete this book.

Dr. Puja Gupta acknowledges the Department of Science and Technology, Government of India, for the Women Scientist-A award (SR/WOS-A/LS-136/2017). She also acknowledges the authorities of RIMT University for providing an opportunity to work as an Assistant Professor in Bioscience.

Dr. Mohd. Shahnawaz also acknowledges the authorities of the University of Ladakh for providing an opportunity to work as an Assistant Professor (Botany) on a contract basis at Kargil Campus.

We also thank our contributing authors for their tireless efforts to submit their quality chapters.

The support tendered by the Commissioning Editor, Dr. Renu Upadhay, and the Handling Editor, Dr. Jyotsna Jangra, is duly acknowledged.

Last, but not least, we are also thankful to our respective parents, better halves and kids for their trust, support and time to complete this book successfully.

Puja Gupta
Gobindgarh, India

Mohd. Shahnawaz (Khakii)
Kargil, India

List of contributors

Rameesha Abid
National Institute of Genomics and Agriculture Biotechnology (NIGAB)
National Agriculture Research Center
Islamabad, Pakistan
and
Department of Microbiology
Quaid-i-Azam University
Islamabad, Pakistan

Somen Acharya
DRDO-Defence Institute of High Altitude Research
Leh-Ladakh, India

Muhammad Adnan
Department of Agriculture
University of Swabi
Swabi, Khyber Pakhtunkhwa, Pakistan

Wajya Ajmal
National Institute of Genomics and Agriculture Biotechnology (NIGAB)
National Agriculture Research Center
Islamabad, Pakistan

Muhammad Yasir Akbar
National Institute of Genomics and Agriculture Biotechnology (NIGAB)
National Agriculture Research Center
Islamabad, Pakistan

Romana Akhtar
Department of Zoology
Govt. Degree College Kishtwar
Kishtwar, Jammu, Jammu and Kashmir, India

Abdel Rahman Al Tawaha
Department of Biological Sciences
Al Hussein Bin Talal University
Ma'an, Jordan

Amjad Ali
Department of Environmental Engineering
Xi'an University of Architecture and Technology
Beilin, China

Amanullah
Department of Agronomy
the University of Agriculture
Peshawar, Pakistan

Osikemekha Anthony Anani
Laboratory for Ecotoxicology and Forensic Biology
Department of Biological Science
Faculty of Science
Edo State University
Uzairue, Edo State, Nigeria

Adnan Ansari
National Center for Bioinformatics
Quaid-i-Azam University
Islamabad, Pakistan

Iqra Asif
Department of Biotechnology
University of Sialkot
Sialkot, Pakistan

Bidyut Bandyopadhyay
Department of Biotechnology
Oriental Institute of Science and Technology
Burdwan, West Bengal, India

Fatima Batool
Department of Botany
Division of Science and Technology
University of Education
Lahore, Pakistan

Neha Rani Bhagat
DRDO-Defence Institute of High Altitude Research
Leh-Ladakh, India

Vijay K. Bharti
DRDO-Defence Institute of High Altitude Research
Leh-Ladakh, India

Neha Bhat
Department of Biotechnology
University of Jammu
Gujarbasti
Jammu, Jammu and Kashmir, India

Rohini Bhat
Interactions Lab
National Centre for Biological Sciences
Bellary Road
Bangalore, Karnataka, India

Koushik Biswas
Department of Agriculture
Brainware University
Barasat
West Bengal, India

Sriya Choudhury
Department of Biotechnology
Oriental Institute of Science and Technology
Burdwan, West Bengal, India

Mudasir Ahmad Dar
Biofuel's Institute
School of the Environment and Safety Engineering
Jiangsu University
Zhenjiang, Jiangsu province, China

Debabrata Das
Department of Agriculture
Brainware University
Barasat
West Bengal, India

Neeraja P. Dhole
Department of Zoology
Savitribai Phule Pune University
Ganeshkhind
Pune Maharashtra, India

Lucinda Elizabeth Doyle
Department of Biochemical Engineering and Biotechnology
Indian Institute of Technology Delhi
New Delhi, India

Shah Fahad
Department of Agronomy
The University Haripur
Khyber Pakhtunkhwa, Pakistan

Rishi Gandhi
Indian Institute of Science Education and Research
Bhopal
Madhya Pradesh, India

Sherish Gaza
Department of Environmental Science
Govt. Degree College Kishtwar
Kishtwar, Jammu, Jammu and Kashmir, India

Shakira Ghazanfar
National Institute of Genomics and Agriculture Biotechnology (NIGAB)
National Agriculture Research Center
Islamabad, Pakistan

Saheli Ghosh
Environmental Biotechnology and Genomics Division
CSIR-National Environmental Engineering Research Institute (NEERI)
Nehru Marg
Nagpur, Maharashtra, India

Honey Goel
Baba Farid University of Health Sciences
Faridkot, Punjab, India

Rakesh Goswami
Department of Biotechnology
Oriental Institute of Science and Technology
Burdwan, West Bengal, India

Ishaan Gupta
Biochemical engineering and Biotechnology department
Indian Institute of Technology Delhi
New Delhi, India

Meenakshi Gupta
Division of Agronomy
Sher-e-Kashmir University of Agricultural Sciences & Technology of Jammu
Chatha
Jammu, Jammu and Kashmir, India

Puja Gupta
Regional Institute of Management and Technology (RIMT)
Mandi Gobindgarh, Sirhind
Punjab, India

Vikas Gupta
Advanced Centre for Rainfed Agriculture
Rakh Dhiansar
Samba, Sher-e-Kashmir University of Agricultural Sciences & Technology of Jammu
Jammu and Kashmir, India

Fariha Hasan
Department of Microbiology
Quaid-i-Azam University
Islamabad, Pakistan

Amara Hassan
Department of Botany
Government College University Faisalabad
Faisalabad, Pakistan

Muhammad Hayat
State Key Laboratory of Microbial Technology
Institute of Microbial Technology
Shandong University
Qingdao, China

Muhammad Ibrar
Key Laboratory of Plant Development and Environment Adaptation Biology
Ministry of Education
School of Life Sciences
Shandong University
Qingdao, China

Jignesh H. Kamdar
Department of Microbiology
School of Science
RK University
Rajkot, Gujarat, India

Hetvi Kanojia
Department of Microbiology
School of Science
RK University
Rajkot, Gujarat, India

Khushal M. Kapadiya
Department of Chemistry
School of Science
RK University
Rajkot, Gujarat, India

Saliha Khalid
Department of Biochemistry
Faculty of Life Science
University of Central Punjab
Lahore, Pakistan

Abeer Khan
Department of Biotechnology
University of Sialkot
Sialkot, Pakistan

Imran Khan
Key Laboratory of Tropical Marine Bio-resources and Ecology
Guangdong Key Laboratory of Marine Materia Medica
South China Sea Institute of Oceanology
Innovation Academy for South China Sea Ecology and Environmental Engineering
Chinese Academy of Sciences
Guangzhou, China

Kiran Kour
Division of Fruit Science
Sher-e-Kashmir University of Agricultural Sciences & Technology of Jammu
Jammu, Jammu and Kashmir, India

Sarabdeep Kour
Division of Soil Science and Agricultural Chemistry
Sher-e-Kashmir University of Agricultural Sciences & Technology of Jammu
Jammu, Jammu and Kashmir, India

Abbhishek Kumar
Biochar Consultant
International Crops Research Institute for Semi-Arid Tropics
Patancheruvu, Telengana, India

Krishna Kumar
DRDO-Defence Institute of High Altitude Research
Leh-Ladakh, India

Sahar Mumtaz
Department of Botany
Division of Science and Technology
University of Education
Lahore, Pakistan

Reem Nadeem
National Center for Bioinformatics
Quaid-i-Azam University
Islamabad, Pakistan

Yashika Nalla
Department of Microbiology
School of Science
RK University
Rajkot, Gujarat, India

Skarma Nonzom
Department of Botany
University of Jammu
Jammu, Jammu and Kashmir, India

Frances Ngozi Olisaka
Department of Environmental Management and Toxicology
Faculty of Life Sciences
University of Benin
Benin City, Edo State, Nigeria

Osayomwanbo Osarenotor
Department of Environmental Management and Toxicology
Faculty of Life Sciences
University of Benin
Benin City, Edo State, Nigeria

Radhakrishna S. Pandit
Department of Zoology
Savitribai Phule Pune University
Ganeshkhind
Pune, Maharashtra, India

Veda Pandya
Department of Microbiology
School of Science
RK University
Rajkot, Gujarat, India

Abida Parveen
Department of Botany
Government College University Faisalabad
Faisalabad, Pakistan

Harsh V. Purohit
Department of Microbiology
School of Science
RK University
Rajkot, Gujarat, India

Asifa Qureshi
Environmental Biotechnology and Genomics Division
CSIR-National Environmental Engineering Research Institute (NEERI)
Nehru Marg
Nagpur, Maharashtra, India

Muhammad Rafiq
Department of Microbiology
Faculty of Life Sciences and Informatics
Balochistan University of I.T. Engineering and Management Sciences
Quetta, Pakistan

Keval Y. Raval
Department of Pharmacology
School of Pharmacy
RK University
Rajkot, Gujarat, India

Benish Sadaqat
Colin Ratledge Center for Microbial Lipids
School of Agriculture Engineering and Food Science
Shandong University of Technology
Zibo, China
and
Division of Biochemistry and Structural Biology
Department of Chemistry
Lund University
Lund, Sweden

Mousumi Saha
E-YUVA Centre (Supported by BIRAC)
Adamas University
West Bengal, India
and
Department of Biotechnology
Oriental Institute of Science and Technology
Burdwan, West Bengal, India

Wasim Sajjad
State Key Laboratory of Cryospheric Science
Northwest Institute of Eco-Environment and Resources
Chinese Academy of Sciences
Lanzhou, China

Manisha K. Sangale
Cyfoeth Naturiol Cymru
Natural Resource Wales
Swansea, United Kingdom

Agniswar Sarkar
Institute of Biomedical Sciences
Affiliated to West Bengal University of Health Science
Chandul, Burdwan East, India

Aqib Sayyed
Department of Biotechnology
College of Life Science and Technology
Huazhong University of Science and Technology
Wuhan, China

Mohd. Shahnawaz
Department of Botany
University of Ladakh
Kargil Campus
Khumbathang, Kargil, Ladakh, India

Richu Singla
Guru Gobind Singh Medical College
Faridkot, Punjab

Mohd. Yaseen Sirwal
Department of Chemistry
Govt. Degree College (PG) Bhadarwah
Bhadarwah, Jammu and Kashmir, India

Geeta Sumbali
Department of Botany
University of Jammu
Jammu, Jammu and Kashmir, India

Fazli Wahid
Department of Agriculture
University of Swabi
Swabi, Khyber Pakhtunkhwa, Pakistan

Harshal Wasnik
Indian Institute of Science Education and Research
Bhopal
Madhya Pradesh, India

Muhammad Zamin
Department of Agriculture
University of Swabi
Swabi, Khyber Pakhtunkhwa, Pakistan

Introduction

Soil contains a vast number of microbes that work cooperatively to hold *nutrients* in the soil. Different edaphic factors are responsible for modifying the soil microbial communities, thereby influencing soil quality and plant health. Various adaptations allow these microbes to cope in different environments, enabling them to survive even in hostile conditions that may not be suitable for human beings to live. One such harsh environment is extremely cold habitats. The adapted microbes produce various biomolecules for adaptation. These biomolecules can be exploited for various pharmaceutical and biotechnological industries.

Experts from different parts of the world have contributed 16 chapters in the current volume as follows:

1. Chapters 1 and 2 overview the physicochemical, biological properties, soil health and biotic communities of soil in cold habitats.
2. Chapters 3 and 4 highlight the impact of soil characteristics on the soil microbial biomass and microbial diversity of five major cold deserts of the world (the Gobi Desert; the Taklamakan Desert, Eurasia; the Namibian Desert, Africa; the Drass Desert, India; and the Atacama Desert, South America).
3. Chapters 5 and 6 discuss the diversity of potassium and iron solubilising bacteria in trans-Himalayan soil and the responses of microfungi to the dynamics of cold environments.
4. Chapters 7 and 8 document the microbial biodiversity of the rhizosphere soils in the Himalayan (Indian part) regions and report the soil microbiome for enhanced crop productivity.
5. Chapters 9 and 10 report the impact of microbial biofilm community shifts on the degradation of microplastics in cold habitats and portray varied adaptive strategies employed by psychrophilic bacteria to live in cold habitats.
6. Chapters 11 and 12 pool the studies on the production of secondary metabolites by some of the important psychrophilic soil microbes to report the important drug molecules and application of 'Omics' technologies to study soil microbiomes.
7. Chapters 13 and 14 describe the ecological role of psychrophiles' adaptation in various cold ecozones and the remediation of soil as a step toward modern soil biotechnology.
8. Chapters 15 and 16 present a new approach to exploring the psychrophilic soil microbiome through machine learning and mass spectrometry as an emerging analytical technique for agriculture in a changing climate.

1 Soil physicochemical and biological properties in cold habitats of the north-west Himalayas

Their constraints and management strategies

Sarabdeep Kour, Meenakshi Gupta, Vikas Gupta and Kiran Kour*

CONTENTS

* Corresponding author E-mail: sarabdeep_kour@rediffmail.com

DOI: 10.1201/9781003354031-1

1.1 INTRODUCTION

Soil is the earth's outermost layer, composed of organic materials, minerals, gases, liquids and creatures that sustain all kinds of life on the planet. There is a great diversity among soil types, and every soil type has specific physicochemical and biological properties that perform all nutrient cycling and make nutrients available for plant growth. Different environments have different soil characteristics. India is reported to have diverse climatic zones extending from tropical (south) to temperate and alpine (north).[1] The northern Himalayan region, often known as the trans-Himalayan belt, is a cold desert environment. It is a harsh, dry, freezing, wind-blown wasteland with temperatures typically staying below 5 °C. The precipitation from the western depression occurs in the form of both rain and snow. Summer temperatures stay over 20 °C, particularly in April and June.[2] The soils of the north-western Himalayan region are becoming degraded and have limited ability to sustainably promote plant growth. Various physical, chemical and biological soil factors are responsible for the degradation of these soils. The most serious threat to land health in these regions is deforestation, resulting in soil erosion, which leads to the loss of nutrient-rich topsoil. Various natural processes, like mineralisation and decomposition, are also reduced to a large extent at lower temperatures, making soil coarse-textured and nutrient-poor in these ecosystems. However, different strategies must be chosen in such regions for maintaining soil health. Nutrient-deficient soils, in turn, affect the soil microbial communities. The microbes found in such environments are armed with adaptations to flourish in the associated stresses. In this chapter, an attempt will be made (a) to review the influence of low temperature on the physicochemical and biological properties of soil in the cold environments of the north-western Himalayas in varied land types, and (b) to interlink the interactions of soil components to manage soil health.

1.2 SOIL PROPERTIES AND THEIR IMPORTANCE

Soil properties are important in sustainable agriculture and can be influenced by crops, fertiliser, management strategies and the period of cultivation.[3] Soil properties include physical, chemical and biological properties.

1.2.1 PHYSICAL PROPERTIES OF SOIL

The physical characteristics give knowledge about soil hydrologic structures such as water entry and withholding, which determines water accessibility to plants. Some characteristics impact rooting volume and aeration status, which are both connected to nutrient availability, whereas some physical properties depict soil erosion status. Typical physical properties include measures of:

a) **Soil structure**

Soil structure explains the organisation of the soil's solid sections and the pore space that exists between them. Water, air mobility, biological activity, root development and seedling emergence are all influenced by soil structure.

b) **Soil texture**

Soil texture refers to the amount of sand, silt and clay in a soil layer. Particles with an equivalent spherical diameter (e.s.d.) of 2 mm are included, since the majority of physicochemical activity occurs in this fine size fraction. This physical feature of soil regulates water and nutrient retention and accessibility to plants.

c) **Aggregate stability**

Soil aggregate stability refers to the capacity of soil aggregates to withstand degradation when subjected to outside factors such as tillage, water, wind erosion, shrinking, and erosion by wind and water. Soil aggregate stability can be influenced by soil management.

d) **Available water holding capacity**

Available water holding capacity (WHC) refers to the quantity of water that soil can store and make accessible to plants. The highest soil water retention capacity for the entire field is referred to as the field capacity.

e) **Bulk density**

Soil bulk density refers to the weight of dry soil per unit of bulk volume, comprising air space. Varied soil types can have significantly different bulk densities and are influenced by management practices. The assimilation of huge amounts of organic matter into the soil will reduce bulk density, but soil compacting procedures will raise bulk density. Mineral soils typically have a bulk density of 1.0 to 1.8 g cm^{-3}.

f) **Soil crusts**

Soil crusts are formed by thick, continuous deposits of soil particles that have not aggregated on the surface of tilled and exposed soils. A soil surface dries up and develops hard structural crusts following irrigation or rainfall. A surface crust results in reduced infiltration, poor seedbed quality, and diminished air exchange between soil and atmosphere.

1.2.2 Chemical properties of soil

The chemical properties of the soil offer details about the balance between soil solution (soil water and nutrients) and exchange sites (clay particles, organic matter), plant health, the plants' dietary needs, the amount of soil contaminants, and their accessibility to animals and plants. The following factors must be studied in depth to comprehend the chemical characteristics of the soil:

a) **Soil pH**

Soil pH depicts whether the soil is acidic or alkaline. Soil pH directly impacts how readily nutrients are available to the plants. Most soils should have a pH between 5.5 and 7.5 for optimal plant development.[4,5]

b) **Electrical conductivity**

Soil electrical conductivity (EC) is the salt content in soil (salinity of soil). It is a crucial factor that indicates soil health. A surplus of salts alters the soil water balance, which prevents plants from growing.[5]

c) **Cation exchange capacity**

Cation exchange capacity (CEC) is a measure of the overall number of cations or the total negative charge that a certain soil can carry. CEC impacts many elements of soil and is used as a measure of soil fertility, indicating the soil's ability to hold onto several minerals (such as K^+, NH_4^+ and Ca^{2+}) in plant-available form.[6]

d) **Available nutrients like nitrogen, phosphorus, potassium, micronutrients, etc.**

Soil is the foremost source of nutrients available to plants for growth. The three essential nutrients are nitrogen (N), phosphorus (P) and potassium (K). Other elements such as calcium, magnesium and sulphur are also vital. Besides, plants also require minute amounts of other elements, known as trace elements, e.g., boron, copper, iron, manganese, molybdenum and zinc. Numerous interconnected factors also affect the amount of nutrients in soil, e.g., particle size, humus content, pH, water content, aeration, temperature, root-surface area and mycorrhizal growth.

e) **Soil organic matter**

Soil organic matter (SOM) is constituted of decomposed plant and animal waste. SOM promotes soil fertility by offering cation exchange sites, being a reservoir of plant nutrients.[4] Therefore, there is a strong relationship between soil fertility and SOM content.

f) **Potential contaminant elements (heavy metals, radioactive compounds, etc.)**

Soil serves as a sink for pollutants. Heavy metals, herbicides, organophosphates and radioactive residues are the principal contaminants carried by soil. These accumulating pollutants damage different living forms and result in a variety of health problems. Industrial processes, agricultural chemicals or inappropriate waste disposal increase these toxins.[7]

1.2.3 Biological Properties of Soil

These properties provide information about micro- and macroorganisms and their activities or end products that support the soil food chain and aid in nutrient cycling and the breakdown of organic materials. These properties include measures of:

a) **Potentially mineralisable nitrogen (PMN)**

PMN is the proportion of organic nitrogen that is converted into forms that are readily available to plants (or minerals) under specific temperature, moisture, aeration and time conditions. Plant and animal tissues, as well as microbial biomass, are the major sources of PMN. It is regarded as an indirect indicator of the availability of nitrogen throughout the growing season for crops, since it indicates the portion of nitrogen that is readily broken down by soil microbes.[8–10]

b) **Microbial biomass C, N, P**

Microbial N is crucial for quantifying the dynamics of N in agricultural ecosystems, as it controls soil organic nitrogen availability and loss, particularly in high-input systems. The dynamics of C and P in soil are also controlled by microbial biomass C and P.[10]

c) **Soil enzymes like dehydrogenase, phosphatase, etc.**

Soil enzymes are utilised as markers of soil quality due to their great sensitivity and quick response to soil management. Additionally, they are simple to assess and evaluate.[11]

1.3 SOIL PROPERTIES AND CHARACTERISTICS OF NORTH-WESTERN HIMALAYAS

The north-western Himalayan region primarily comprises the union territories of Jammu & Kashmir and Ladakh (erstwhile J&K state), Himachal Pradesh and Uttarakhand. This tract of the Himalayan system is located between 28°43′ N to 37°05′ N latitude and 72°02′ E to 81°02′ E longitude. This trail spans an area of approximately 33.12 Mha. The Himalayas have a significant impact on the climate in northern India. These areas experience a range of climates, including cold, desert zones, temperate, subtropical transitional zones and low-altitude subtropical zones.[12] The climate has a great influence on the physicochemical characteristics of soil, e.g., soil moisture, soil texture, soil pH, soil available macro- and micronutrient status, soil respiration, soil microbial biomass, etc. The north-western ecosystem includes an extensive range of glaciers, hot springs and cold deserts. This tract is broadly divided into three distinct landform units: Greater Himalayas, Lesser Himalayas and Shivaliks.

Nowadays, soils of the north-western Himalayan region are highly degraded, like the north-eastern hill regions,[13] with limited ability to support plant growth. There are various factors responsible for degradation in these soils, e.g., physical (deterioration in soil physical properties), chemical (depletion in soil nutrient status) and biological degradation (less organic matter in the soil, less soil biomass carbon, etc.). The untimely rains/snow, freezing temperatures and melting of snow due to changing climatic conditions make the environment more adverse. Such variations in a microenvironment lead to a change in the properties of the ecosystem. Because of the lower temperatures, various natural processes like mineralisation and decomposition are greatly affected, resulting in lower values of available nutrients. Significant changes in soil organic carbon (SOC) over the altitudinal gradient might be responsible for lower temperature and increasing precipitation with altitude, which results in decreased organic matter decomposition with an increase in altitude.[14] These habitats have nutrient-poor, coarse-textured soil.

1.4 SOIL PHYSICOCHEMICAL PROPERTIES IN COLD ENVIRONMENTS OF NORTH-WESTERN HIMALAYAS

The chemical characteristics of soil, e.g., organic carbon content and soil pH, are the important chemical attributes that in turn affect the availability of nutrients, the quantity of salts present, cation exchange capacity, etc. Chemical properties vary according to the range of altitude, e.g., soils of low and intermediate altitudes are neutral to slightly alkaline, while soils of higher altitudes are acidic.[15] Kumar et al.[16] assessed the cold arid zone soils under diverse land uses in the north-western Himalayas of Himachal Pradesh and reported that forest soils contain higher organic carbon content than cultivated soils and pasture lands. They further reported that both surface (0–15 cm) and subsurface (15–30 cm) soils of cultivated and pasture lands showed a slightly alkaline pH range, varying between 7.7 and 8.0. The textural class of these soils varies from sandy loam to sandy clay loam. The water holding capacity of pasture land is reported to be low as compared with cultivated lands, and this may be due to higher clay and organic carbon content.[16] Vaidya et al.[17] analysed the soil quality of mid-hill regions of the north-western Himalayas in Solan and Kullu districts of Himachal Pradesh under different land-use systems, like traditional agriculture, commercial vegetable cultivation, orchard, forest, and urban, and they reported that all the soil properties showed a similar trend to those of temperate regions. Soils under forest land-use systems registered higher organic carbon (22.43 g kg^{-1}), microbial biomass carbon (120.40 g kg^{-1}) and WHC. In contrast, these soil properties showed lower values under urban land use, whereas higher available nitrogen (N), phosphorus (P), potassium (K), copper (Cu) and iron (Fe) content was found in the vegetable land-use system.[17] Kaur et al.[23] studied soil physical and chemical properties in the mid-hill region of the Jammu division and reported wide variation in soil texture from sandy loam to clay loam. Organic carbon content varied from low to high (2.0 to 9.6 g kg^{-1}), which decreased with increasing depth, whereas soil pH and EC ranged from 5.9 to 8.4 and 0.03 to 1.45 dS m^{-1}, respectively. They also reported that exchangeable aluminium and iron oxides increased with increasing depth.[18]

Gahalod et al.[19] worked on the soil organic carbon stock in 0–25-cm layers of three different altitudinal gradients, <1250, 1250–2500 and >2500 m, under different land-use systems (forests, agriculture, plantations, and grassland/open scrub) in Uttarakhand state. According to their findings, forests and grasslands hold more SOC stock, ranging from 51 to 160 t C ha^{-1}, in comparison to the soils under agricultural land use, and SOC stock also increased with an increase in altitude.[19] Jehangir et al.[20] assessed the soil quality of clear-cut and adjacent intact forests in the coniferous forest of Kashmir. It was revealed that clear-cut soils showed significantly higher WHC, organic carbon, soil temperature, exchangeable calcium, exchangeable potassium and moisture content than intact forests. The increased nutrient content in the clear-cut site could be due to the herbaceous vegetation and stump roots left at that site aiding in maintaining soil fertility.[20] Further, when two leading forest types in the western Himalayas, pine (*Pinus roxburghii*) and oak (*Quercus leucotrichophora*), were compared, oak forests, in general, showed significantly higher physicochemical properties in comparison to pine forests. Across the three soil depths, all the soil parameters, i.e., WHC, organic carbon, total nitrogen and soil moisture, were substantially reduced with growing depth in both forests. However, pH did not display any pattern across the three soil depths in either of the forest types. The surface layer (0–30 cm depth) had an elevated concentration of SOC and total nitrogen in both forests. Deep-rooted oak forests have a competitive advantage over shallow-rooted pine forests in terms of nutrient extraction. Thus, oak forests should be prioritised above pine forests to absorb and retain increased levels of carbon in the soil pool in afforestation and conservation efforts.[21] However, in submontane areas of this region, the physical attributes, i.e., particle density and bulk density, were greater in degraded lands, whereas water storage capacity was higher in forest lands and minimal in degraded lands in both surface and subsurface layers.[22] These soil physical properties, i.e., particle density, soil porosity, bulk density and organic carbon content, were good indices for predicting soil structural stability (Table 1.1).

However, in this region's submontane zones, physical attributes such as bulk density and particle density were greater in deteriorated lands, whereas the maximum WHC was higher in forest lands and lower in deteriorated lands in both the surface and subsurface layers.[22] Soil porosity, bulk density, particle density and

TABLE 1.1
Physical properties of soil under different land-use systems in the submontane region of the north-western Himalayas

Land use	Bulk density (Mg m^{-3})	Particle density (Mg m^{-3})	Porosity (%)	MWHC (%)
Top soil (0–15 cm)				
Forest	1.33	2.52	49.96	41.81
Horticulture	1.49	2.54	43.29	37.37
Agriculture	1.52	2.59	42.56	35.00
Pasture	1.32	2.40	50.28	39.78
Degraded	1.54	2.64	41.74	29.83
LSD (0.05)	0.08	0.07	3.37	05.51
Soil layer (15–30 cm)				
Forest	0.35	2.54	48.82	40.73
Horticulture	1.51	2.57	42.91	36.38
Agriculture	1.54	2.61	41.83	34.13
Pasture	1.34	2.42	49.35	38.06
Degraded	1.57	2.66	40.86	28.90
LSD (0.05)	0.08	0.07	03.39	05.31

Source: From Kumar et al., *JSWC*, 17, 117, 2018. With permission.[22]

organic carbon content were all effective predictors of soil structural stability (Table 1.1).

Kour and Jalali[18] made a comparative study of top soils (0–15 cm) of the three agro-climatic zones of the Jammu region, i.e., subtropical, intermediate and temperate regions, and revealed that soils were mostly acidic in the temperate zone and neutral to slightly alkaline in the intermediate and subtropical zones. Organic carbon content was also higher in the temperate zone, varying from 9.0 to 10.8 g kg^{-1} in these soils, whereas it was lower in subtropical soils, varying from 3.0 to 6.0 g kg^{-1} with a mean value of 4.1 g kg^{-1} (Table 1.2). Clay percentage was also higher in temperate and intermediate zone soils, with mean values of 24.14% and 26.84%, respectively, in comparison to subtropical soils, with a mean value of 15.99%, which could be on account of elevated organic carbon content in these regions. Cation exchange capacity, exchangeable Al and free iron oxides were also higher in temperate and intermediate soils in comparison to subtropical soils.[23]

Chhina et al.[24] reported deforestation on a large scale in the Shivalik region of north-western India, resulting in the conversion of wild forests into farmed lands, which leads to various soil physical and chemical properties being affected. Soil structural characteristics in forest land-use systems, i.e., maize-wheat, agri-horticulture and agri-forestry, were influenced significantly, as shown by the soil down to the depth of 120 cm under the agroforestry system recording significantly elevated soil organic carbon content in comparison to the soil under other land uses, whereas soil physical properties like bulk density were higher (1.57 Mg m^{-3}) under maize-wheat and lower in soils under an agroforestry system for 3 years (1.41 Mg m^{-3}). The overall permeability was lowest in maize-wheat soils as compared to other land uses. Soils under agroforestry planting had greater mean weight diameter and water-stable macro-aggregates after 6 years than soils under the maize-wheat system. Water-stable aggregates, on the other hand, were lower in soils under a 6-year agroforestry crop.[24] So, for maintaining soil health and sustaining agriculture productivity in the north-western Himalayas, different strategies should be adopted. One of these should be conservation agriculture practices. Conservation agriculture (CA)-based tillage strategies such as zero tillage (ZT) and permanent raised beds (PBs) have the potential to improve various soil physical, chemical and biological properties for preserving soil health in agricultural fields of the Indo-Gangetic plains of the north-western Himalayas. According to Parihar and colleagues,[25] physical qualities such as water-stable aggregates were greater, although bulk density and penetration resistance decreased significantly, in ZT and PBs in the 0–15 and 15–30-cm soil layers. Besides, organic carbon, a crucial chemical property regulating nutrient supply in the soil, showed an increase in both the soil layers with CA practices.[25] Soil biological characteristics are impacted by physical and chemical soil factors, which assist in assessing soil quality. Soil quality can be quantified by indicators that can measure slight changes in pH.[26,27] Soil pH, for example, regulates microbial activity as well as microbe and plant nutrient availability. Organic carbon is necessary for soil fertility because it serves as a substrate for bacteria as well as a source of energy, nitrogen, sulphur and other mineral nutrients.[28]

TABLE 1.2
Soil (surface 0–15 cm) physicochemical properties of representative locations of three agro-climatic zones of the Jammu region

Soil property	Subtropical zone		Intermediate zone		Temperate zone	
	Range	Mean	Range	Mean	Range	Mean
pH (1:2.5)	07.40–08.60	7.5	06.50–08.4.0	7.9	06.70–07.20	7.1
EC (dSm^{-1})	00.11–00.62	00.25	00.05–00.42	00.14	00.11–00.19	00.15
OC (g kg^{-1})	03.00–06.00	04.10	04.20–09.60	06.70	09.00–10.80	10.00
$CaCO_3$ (%)	00.60–01.58	01.13	00.30–01.72	01.06	00.55–00.65	00.60
Sand (%)	45.64–85.12	62.45	19.00–62.70	38.51	45.12–52.60	48.10
Silt (%)	07.76–34.18	21.57	20.20–46.90	34.54	25.05–31.00	27.76
Clay (%)	06.86–21.88	15.99	17.00–40.20	26.84	22.35–26.88	24.14
Cation exchange capacity (cmol(p+)kg^{-1})	07.80–10.80	09.20	08.00–13.80	11.60	14.00–14.90	14.40
Exch. Al (mg kg^{-1})	02.40–04.70	03.40	02.70–10.40	05.20	04.70–08.30	06.40
Free Fe oxides (%)	00.31–00.61	00.45	00.36–01.32	00.81	00.69–01.22	00.93
Available sulphur (mg kg^{-1})	08.00–34.50	20.40	09.60–51.10	29.60	10.00–35.70	23.80

Source: adapted from Kour, S. and Jalali, V.K., *J. Indian Soc. Soil Sci.*, 56, 309, 2008.

1.5 SOIL BIOLOGICAL PROPERTIES IN THE NORTH-WESTERN HIMALAYAS

Microbial communities are an integral part of the ecosystem and are very sensitive to environmental conditions. Changing land use affects soil organic carbon, microbial biomass carbon and as a result, the makeup of the soil microbial community.[29] Enzymatic activity is an essential biological indicator of soil quality among all microbial features. Surface soil dehydrogenase activities are lower on desolate land than in forest and cultivated soils.[29] Debnath et al.[30] reported soil biological properties (e.g., soil microbial biomass carbon, soil microbial biomass N, soil microbial P, and enzyme [dehydrogenase, phosphomonoesterase (acid and alkaline) and urease] activities) in the soil at two different depths (0–20 and 21–40 cm) in fruit orchards of different crops soil of peach, soil of plum, soil of apricot and soil of cherry in the Kashmir valley. Unlike microbial biomass, peach had the greatest surface layer dehydrogenase activity (318 μg TPF g $soil^{-1}$ h^{-1}), and the control plot had the lowest (166 μg TPF g $soil^{-1}$ h^{-1}). In the surface soil, peach had the highest alkaline phosphomonoesterase activity (318 μg TPF g^{-1} soil $24h^{-1}$), while apricot had the highest acid phosphomonoesterase (306 μg PNP g $soil^{-1}$ h^{-1}). The Himalayan regions are reported to be strongly influenced by the type of fruit crops, and these decline with increasing soil depth.[30] Soil enzymes were found in land-use systems of comparable soil types at depths ranging from 0 to 15 cm (dehydrogenase, acid phosphatase, alkaline phosphatase, nitrate reductase, arylsulphatase and phytase). Oak (*Quercus incana*), deodar (*Cedrus deodara*), pine (*Pinus roxburghii*), apple orchards and crop-based systems in uplands and valleys were among the land-use systems studied. Tree-based cropping systems increased soil quality by considerably improving all soil microbiological parameters in agroforestry and agri-horticultural systems when compared with cereal-cereal cropping systems of maize-wheat.[31]

The forest habitat has much greater soil enzyme levels than an organic farm or a standard maize-wheat farm soil.[32] Devetter and colleagues[33] investigated the effects of elevation and soil chemistry on the diversity and distribution of rotifers, nematodes and tardigrades along two altitudinal gradients in the north-west Himalayas (Zanskar transect from 3805 to 4714 m.a.s.l) and southern Tibet (Tso Moriri transect from 4477 to 6176 m.a.s.l). The Zanskar transect had higher precipitation, soil moisture, organic matter and accessible nutrients than the arid Tso Moriri transect. In total, 40 nematode species were found. Soil nitrogen, phosphorus and organic matter were shown to be connected to the diversity and dispersion of soil microfauna. More diversified and plentiful faunal groups were found on richer soils.[33]

Among the strategies to maintain soil biological health in the north-western Himalayas, CA practices showed good results. CA-based tillage methods such as ZT and PB increased soil biological qualities for preserving soil health in agriculture fields of the Indo-Gangetic plains of the north-western Himalayas. Soil microbial biomass carbon (MBC) rose by 45–48.9% in the 0–30-cm profile depth of sandy loam soil under a CA-based system. Soil enzymatic activity increased as well, with CA-based treatments showing a considerable improvement in soil enzymes such as fluorescein diacetate, dehydrogenase and alkaline phosphatase.[25] MBC increased with rising fertiliser dose due to enhanced root development, which eventually leads to greater root exudates. For 20 years, continuous application of organics, such as farmyard manure, wheat straw or green manure, in conjunction with inorganics resulted in a large increase in MBC.[34] Besides crops, microorganism species also differ in these cold and high-altitude environments. There is a great diversity in the microbial population in cold climate soils, and the soil bacterial communities are the most complex. Psychrophiles and psychrotrophs are mainly found in elevated altitudes with low temperatures.[35–38] Such microbes have adapted to flourish under these harsh and unfavourable conditions.[39]

1.6 SOIL HEALTH CONSTRAINTS IN NORTH-WESTERN HIMALAYAS

Soils all over the world are subjected to various kinds of degradation, which is a serious global issue, as it results in poor soil health. It might be difficult to remediate soil that has been damaged or polluted.[40] The north-western Himalayan states face land degradation due to deforestation, which results in surface runoff.[41] The soils are susceptible to runoff due to their existence on mountainous terrain. Soil erosion reached 168 tons per hectare per year in the Shivalik ranges of Uttarakhand, Himachal and J&K, as well as Punjab and Haryana.[42] Landslides, poor land use, overgrazing of pastures, uncontrolled development of roads/buildings, mountain blasting and other causes are also responsible for soil erosion.[43]

1.7 STRATEGIES FOR MANAGEMENT OF SOIL HEALTH IN THE NORTH-WESTERN HIMALAYAS

Land degradation affects around 31.6% of the entire geographical area of J&K state due to, e.g., water and wind erosion, and physical and chemical degradation.[40] Thus, technological innovations are urgently needed to overcome land degradation and maintain soil health. Strategies for maintaining soil health include:

- Land levelling and terrace management in sloping lands.
- Bunding, land shaping, re-vegetation, terracing, etc.
- Integrated watershed development to further prevent soil deterioration, runoff and soil erosion.
- Cultivation of forages and legumes, and rotational grazing.
- Soil and water conservation measures.
- Judicious usage of chemical fertilisers.

1.8 SUMMARY AND CONCLUSIONS

Soil quality is determined by physical, chemical and biological qualities, which are influenced by management and land-use type. However, soil quality is degrading day by day due to faulty agricultural practices and overexploitation. Land degradation is caused by farming in fragile deserts, marginal slope lands with no conservation measures, land clearance by deforestation, exhaustion of soil nutrients due to poor agricultural methods, livestock grazing, more irrigation, over-drafting (the process of taking groundwater above the aquifer's safe supply), commercial development and land pollution, including industrial waste dumping on arable lands. The soils in cold habitats of the north-western Himalayas are additionally limited by variables such as steep slopes, soil erosion, nutrient deficits and so on. However, CA combined with additional technologies such as micro-irrigation, terrace management, problem soil management, crop residue recycling, and soil and water conservation methods shows tremendous potential for increasing crop and fruit yield and reversing soil degradation.

ACKNOWLEDGEMENT

All the authors are thankful to the Director Research, Sher-e-Kashmir University of Agricultural Sciences & Technology of Jammu, Chatha for the constant support, inspiration, and providing infrastructural facilities.

REFERENCES

1. Yaduvanshi, Aradhana, Tiro Nkemelang, Ramkumar Bendapudi et al. "Temperature and rainfall extremes change under current and future global warming levels across Indian climate zones." *Weather and Climate Extremes* 31 (2021): 100291.
2. Roy, Arnab, and Uday Kumar. "Emerging farming systems in Western Himalaya: A state level analysis of sustainability." *International Journal for Environmental Sciences and Natural Resources* 9, no. 2 (2018): 555757.
3. Qian, Xun, Jie Gu, Wei Sun et al. "Changes in the soil nutrient levels, enzyme activities, microbial community function, and structure during apple orchard maturation." *Applied Soil Ecology* 77 (2014): 18–25.
4. Brady, N. C., and R. R. Well. "The nature and properties of soils Prentice Hill, Upper Sadle River." New Jersey (1999).
5. Minasny, Budiman, Suk Young Hong, Alfred E. Hartemink et al. "Soil pH increase under paddy in South Korea between 2000 and 2012." *Agriculture, Ecosystems & Environment* 221 (2016): 205–213.
6. Birkeland, Peter W. *Soils and Geomorphology.* Oxford University Press, 1984.
7. Ashraf, Muhammad Aqeel, Mohd Jamil Maah, and Ismail Yusoff. "Soil contamination, risk assessment and remediation." *Environmental Risk Assessment of Soil Contamination* 1 (2014): 3–56.
8. Doran, John W., and Timothy B. Parkin. "Defining and assessing soil quality." *Defining Soil Quality for a Sustainable Environment* 35 (1994): 1–21.
9. Moore, J. M., Susanne Klose, and M. A. Tabatabai. "Soil microbial biomass carbon and nitrogen as affected by cropping systems." *Biology and Fertility of Soils* 31, no. 3 (2000): 200–210.

10. Clark, Jason D., Fabián G. Fernández, Kristen S. Veum et al. "Soil-nitrogen, potentially mineralizable-nitrogen, and field condition information marginally improves corn nitrogen management." *Agronomy Journal* 112, no. 5 (2020): 4332–4343.
11. Wolinska, A., Magdalena Zapasek, and Z. Stepniewska. "The optimal TTC dose and its chemical reduction level during soil dehydrogenase activity assay." *Acta Agrophysica* 23, no. 2 (2016).
12. Sidhu, G. S., and Jaya N. Surya. "Soils of North-Western Himalayan eco-system and their land use, constraints, productivity potentials and future strategies." *Agropedology* 24, no. 1 (2014): 1–19.
13. Gupta, R. D., and P. D. Sharma. "Soil erosion and conservation in the northwest Himalayas". *Natural Resource Management for Sustainable Hill Agriculture* 2 (2007): 205–214.
14. Araujo, Marla A., Yuri L. Zinn, and Rattan Lal. "Soil parent material, texture and oxide contents have little effect on soil organic carbon retention in tropical highlands." *Geoderma* 300 (2017): 1–10.
15. Lone, Parvaiz Ahmad, Ajay Kumar Bhardwaj et al. "Assessment of soil macronutrient status of some threatened medicinal plants of Kashmir Himalaya, India." *Research Journal of Botany* 11, no. 1/3 (2016): 18–24.
16. Kumar, Dhirender, Daulat Ram Bhardwaj, Prashant Sharma et al. "Population dynamics of Juniperus macropoda Bossier forest ecosystem in relation to soil physico-chemical characteristics in the cold desert of North-Western Himalaya." *Forests* 13, no. 10 (2022): 1624.
17. Vaidya, Pratima, Satish Kumar Bhardwaj, Sumita Chandel et al. "Impact of land uses on soil quality in mid hill region of North Western Himalayas." *The Ecoscan* 10 (2016): 553–559.
18. Kour, Sarabdeep, and V. K. Jalali. "Forms of sulphur and their relationships in soils of different agro-climatic zones of Jammu region." *Journal of the Indian Society of Soil Science* 56, no. 3 (2008): 309–312.
19. Gahalod, N. S., Navneet Jaryal, Mallikarjun Roodagi et al. "Influence of altitude on soil organic carbon stocks in land use systems of Uttarakhand State: Geospatial approach." *Journal of Soil and Water Conservation* 17, no. 4 (2018): 311–316.
20. Jehangir, Arshid, A. R. Yousuf, Z. A. Reshi et al. "Comparison of physical, chemical and microbial properties of soils in a clear-cut and adjacent intact forest in North Western Himalaya, India." *International Journal of Soil Science* 7, no. 3 (2012): 71.
21. Joshi, Gunjan, and G. C. S. Negi. "Physico-chemical properties along soil profiles of two dominant forest types in Western Himalaya." *Current Science* (2015): 798–803.
22. Kumar, Vijay, K. R. Sharma, Vivak M. Arya, et al. "Landuse effects on structural stability and soil organic carbon in the submontane areas of north-western Himalayas, India." *Journal of Soil and Water Conservation* 17, no. 2 (2018): 117–122.
23. Kour, Sarabdeep, Sanjay Arora, V. K. Jalali, and A. K. Mondal. "Soil sulfur forms in relation to physical and chemical properties of midhill soils of North India." *Communications in Soil Science and Plant Analysis* 41, no. 3 (2010): 277–289.
24. Chhina, Lovedeep Kaur, R. K. Gupta, and S. S. Kukal. "Soil structural attributes in relation to land use in Shiwaliks region of northwest India." *Journal of Soil and Water Conservation* 18, no. 1 (2019): 22–26.
25. Parihar, Chitar Mal, M. R. Yadav, Shankar L. Jat et al. "Long term effect of conservation agriculture in maize rotations on total organic carbon, physical and biological properties of a sandy loam soil in north-western Indo-Gangetic Plains." *Soil and Tillage Research* 161 (2016): 116–128.

26. Doran, J. W. "Microbial biomass and mineralizable nitrogen distributions in no-tillage and plowed soils." *Biology and Fertility of Soils* 5, no. 1 (1987): 68–75.
27. Baretta, Dilmar, George Gardner Brown, and Elke Jurandy Bran Nogueira Cardoso. "Potencial de la macrofauna y de otras variables edáficas como indicadoras de la calidad del suelo en áreas con Araucaria angustifolia." *Acta zoológica mexicana* 26, no. SPE2 (2010): 135–150.
28. Idowu, Omololu J., Harold M. Van Es, George S. Abawi et al. "Farmer-oriented assessment of soil quality using field, laboratory, and VNIR spectroscopy methods." *Plant and Soil* 307, no. 1 (2008): 243–253.
29. Arunkumar, K., R. D. Singh, A. K. Patra et al. "Probing of microbial community structure, dehydrogenase and soil carbon in-relation to different land uses in soils of Ranichauri (Garhwal Himalayas)." *Int. J. Curr. Microbiol. App. Sci* 2, no. 11 (2013): 325–338.
30. Debnath, S., A. K. Patra, N. Ahmed et al. "Assessment of microbial biomass and enzyme activities in soil under temperate fruit crops in north western himalayan region." *Journal of Soil Science and Plant Nutrition* 15, no. 4 (2015): 848–866.
31. Dhaliwal, Jeevanjot, S. S. Kukal, and Sandeep Sharma. "Improving soil microbiological properties in typic ustochrepts through tree-based cropping system." *Journal of the Indian Society of Soil Science* 66, no. 2 (2018): 222–228.
32. Singh, Raj Deo, Arun Kumar, Ashok Kumar Patra et al. "Impact of different land use management on soil enzyme activities and bacterial genetic fingerprints of North-Western Himalayas." *Current World Environment* 9, no. 3 (2014): 728.
33. Devetter, Miloslav, Ladislav Háněl, Klára Řeháková et al. "Diversity and feeding strategies of soil microfauna along elevation gradients in Himalayan cold deserts." *PloS One* 12, no. 11 (2017): e0187646.
34. Sharma, Upinder, and S. K. Subehia. "Effect of long-term integrated nutrient management on rice (Oryza sativa L.)-wheat (*Triticum aestivum* L.) productivity and soil properties in North-Western Himalaya." *Journal of the Indian Society of Soil Science* 62, no. 3 (2014): 248–254.
35. Ghildiyal, A., and A. Pandey. "Isolation of cold tolerant antifungal strains of Trichoderma sp. from glacial sites of Indian Himalayan region." *Research Journal of Microbiology* 3, no. 8 (2008): 559–564.
36. Pandey, Anita, Namrata Das, Bhavesh Kumar et al. "Phosphate solubilization by Penicillium spp. isolated from soil samples of Indian Himalayan region." *World Journal of Microbiology and Biotechnology* 24, no. 1 (2008): 97–102.
37. Sati, Priyanka, Kusum Dhakar, and Anita Pandey. "Microbial diversity in soil under potato cultivation from cold desert Himalaya, India." *International Scholarly Research Notices* 2013 (2013).
38. Yadav, Ajar Nath, Shashwati Ghosh Sachan, Priyanka Verma et al. "Prospecting cold deserts of north western Himalayas for microbial diversity and plant growth promoting attributes." *Journal of Bioscience and Bioengineering* 119, no. 6 (2015): 683–693.
39. Moyer, Craig L., and Richard Y. Morita. "Psychrophiles and psychrotrophs." *Encyclopedia of Life Sciences* 1, no. 6 (2007).
40. Bhattacharyya, Ranjan, Birendra Nath Ghosh, Prasanta Kumar Mishra et al. "Soil degradation in India: Challenges and potential solutions." *Sustainability* 7, no. 4 (2015): 3528–3570.
41. Arora, Sanjay, Vikas Sharma, A. Kohli et al. "Soil and water conservation for sustainable crop production in Kandi region of Jammu." *Journal of Soil and Water Conservation, India* 5, no 2 (2022): 77–82.

42. Sharma, P. D. "Managing natural resources in the Indian Himalayas." *Journal of the Indian Society of Soil Science* 52, no. 4 (2004): 314–331.
43. Gupta, R. D., Gupta, J. P., and Singh, H. "Problems of agriculture in Kashmir Himalayas." In: *Conserving Indian Environment*, edited by Chadha, S. K. Pointer's Publishers, 1992.

2 Soil as a host to the biotic community

Harsh V. Purohit, Hetvi Kanojia, Veda Pandya, Yashika Nalla, Keval Y. Raval, Khushal M. Kapadiya and Jignesh H. Kamdar**

CONTENTS

* Corresponding author e-mail: khushal.kapadiya@rku.ac.in; jignesh.kamdar@rku.ac.in

DOI: 10.1201/9781003354031-2

2.1 INTRODUCTION

Soil, the planet's largest biodiversity depository, is one of the most complex and diverse biotic communities. It has many microorganisms (MOs) (bacteria, fungi, etc.) and higher organisms (arthropods, mites). More research shows that various factors affect soil range and composition. Root MOs may help plants create resistance habitats for abiotic tolerance.[1] In the current chapter, an attempt has been made to overview soil microbial diversity, relationships, food systems, soil habitats, variables and soil biotic community responsibilities.

2.2 SOIL MICROORGANISM DIVERSITY

2.2.1 Bacteria

After 100 years of investigation, most soil bacterium varieties are still unknown. Previous studies found that few soil samples share bacteria. Bacteria in soils vary. Some soil bacterial taxa are more common than others. Bradyrhizobium dominates North American forest soils.[2] Soil microbes have been essential to agricultural biogeochemical cycles for decades. Plant–bacteria interactions in the rhizosphere impact plant health and soil potency.[3] Thus, understanding soil's numerous and ubiquitous bacterial populations is crucial.

2.2.2 Fungi

Fungi are becoming important ecosystem regulators. They are vital to soil fertility. In high-fertility soils, multiple fungal species interact.[4] Most of the 80,000+ fungal species have been described.[5] Fungi range from microscopic yeasts to huge macrofungi, toadstools and giant puffballs. Fungi impact several soil microbiological activities, including soil fertility, plant and animal matter breakdown, chemical recycling, and plant resistance and health. Fungi are structurally and functionally diverse and adopt different strategies as saprophytes, symbionts or pathogens.[6] Fungi mostly destroy soil. Soil fungi communicate and perform various tasks. Mycorrhizal linkages can help seedlings share between plants or defend against infections. Arbuscular mycorrhizal fungi (AFM) are the most common in agricultural soils. They comprise 5–50% of all soil microbial biomass. The AFM association's role in agricultural sustainability has garnered recognition. Most tropical crops are mycorrhizal and respond significantly to AMF. They control soil health and fertility. Additionally, soil fungi interact with plants, arthropods, nematodes and pathogenic fungi. Invertebrates and toxic microbes may eat them. Soil fungi may be active or dormant. Thus, soil fungi comprise a variety of organisms that may be inactive, actively developing, or intimately related to other species or plants.

2.2.3 Soil fauna

How organisms exploit soil or habitat defines soil fauna. Ladybird beetles, which dwell on plants and hibernate in soil, are transitory species. Night-time cutworm larvae eat seedlings. Nematodes feed on insects to complete their life cycle. Velvet mites are seasonal residents that live underground and breed as adults. Collembolans are soil-dwelling microarthropods. A length-based classification divides soil fauna into four sizes: micro, meso, macro and mega. Water film microfauna include protozoa, rotifers, tardigrades and nematodes. Mesofauna inhabit air-filled pores. Megafauna considerably affect gross soil structure, whereas macrofauna may scrabble out their own territories.[7]

2.2.4 Microfauna

Rhizospheric microfauna are critical for plant nutrition release, soil organic carbon buildup, multi-trophic interactions above ground and polluted soil bioremediation. The earth, above-ground community and rhizosphere biological activity benefit from microfauna. The most common soil microfauna include animal-like protists, nematodes and rotifers. Heterotrophic microfauna ("unicellular creatures that live in soil water films") are animal-like protists and eat bacteria, although some are predatory fungivores. Soil contains several nematodes (roundworms). Free-living and plant-parasitic nematodes dwell in soil water films and plant roots. Common soil nematodes are "Rhabditida," "Tylenchida," "Aphelenchida" and "Dorylaimida," based on their eating habits.[8]

2.2.4.1 Protozoa

Sarcomastigophora and Ciliophora are soil and litter protozoa. Flagellates, naked amoebae, Testacea and ciliates are ecological groups. Flagellates have whip-like flagella and are the most numerous and active protozoa. They depend on bacteria for nutrient cycling. Desert soils had 102–105 g^{-1} values.[9] Amoebae, insatiable soil protozoa, inhabit agricultural, grassland and wooded soils. Amoebae and larger species like ciliates eat phagotrophically. Testate amoebae are rarer than naked ones except in wet, forested areas. Ciliates have unusual life cycles and complex generating patterns. They outnumber other categories. Ciliates feed on microorganisms in soil pores and fissures. Refractory ciliates appear in food-rich environments.

2.2.4.2 Rotifera

Water film soil usually includes these tiny animals. Rotifera are 0.05–3 mm long and not listed in soil biota compendia. Dry Antarctic Valley soils contain rotifers. Rotifers are distinguished by their pseudo-segmented body wall (lorica), "crown" with cilia, and feeding mechanism with strong muscles and jaws. Nine out of ten soil rotifers are Bdelloidea. Besides "biogeochemical cycles," their importance is uncertain.[10]

2.2.5 Mesofauna

Mesofauna occupy air-filled pores, primarily existing ones. Soil mesofauna taxa, commonly known as microarthropods, include terrestrial Aptera. Acari, Collembolans, Proturans, Diplurans, Symphellids and Enchytraeids are also included. Microarthropods link microfauna and microorganisms to mesofauna, since they feed on nematodes and fungi.[11]

2.2.5.1 Nematoda

Soil nematodes are multicellular roundworms. Four out of every five organisms are nematodes. As below-ground parasites and plant pathogens, nematodes regulate decomposition and NPP. Nematodes are "tubes within tubes" that devour a variety of meals. They consume bacteria, fungi, plant feeders, omnivores and predators.[12]

2.2.5.2 Microarthropods

Microarthropods populate most soils (primarily mites and Collembolans) and greatly impact forest floor decomposition. Many microarthropods eat fungi and nematodes, linking them to the mesofauna and microorganisms. Scavengers on soil nematodes include eight "oribatid mites" and one "mesostigmatid."[13]

2.2.5.3 Enchytraeids

"Terrestrial Oligochaeta" Enchytraeidae will be discussed under Macrofauna. Over 700 "microdrile" oligochaete species are little, non-pigmented "pot worms." Enchytraeidae inhabit damp, organic-rich woodland soils. Enchytraeids are now found worldwide.[14] Enchytraeids eat mineral and organic particles, although less than earthworms. Numerous investigations have found that enchytraeid diets are rich in finely divided plant components and bacterial and fungal hyphae. Enchytraeids live in damp soils from the Arctic to the tropics.[15]

2.2.6 Macrofauna

Macrofauna – arthropods, molluscs and earthworms – are visible to the naked eye (4 mm × 2 mm). Epigeic, anecic and endogeic subcategories contribute differentially to the environment. Epigeics, small, pigmented earthworms and saprophagous arthropods, feed on surface trash. The group's predators are coleopterans. Anecics eat surface garbage yet create underground grubs for protection. They include most termites and certain large, coloured earthworms. Endogeics – termites and colourless earthworms – comprise most soil organisms.

2.2.6.1 Macroarthropods

"Macroarthropods" include enormous insects, spiders, myriapods and others. Terrestrial macroarthropods live above and below ground. Several "cryptozoan" macroarthropods live in cracks, fissures, bark, stones and logs. Cryptozoans feed at night and are little understood. Soil ecosystems and food webs rely on macroarthropods, which can directly alter soil structure.

2.2.6.2 Oligochaeta (Earthworms)

Earthworms are vital soil fauna. Earthworms comprise Oligochaeta and Annelida. Ecologically, Lumbricidae and Megascolecidae are most important. Humans introduced several of these species, which now dominate temperate earthworm biodiversity. Deserts, tundra and polar environments have the lowest concentrations and temperate, tropical and grasslands the highest.

2.2.6.3 Formicidae (Ants)

Formicidae (ants) are the most important soil insects because of their huge impact on soil structure. Ants are abundant and diverse in Arctic and tropical habitats. Ant communities include many species, including tropical species. Ants live in multi-caste colonies. They greatly affect their habitats. They are predators of small invertebrates, including predators like carabid beetles and spiders.[16] Ants, like earthworms, build ecosystems by moving soil. Ants affect soil structure most in deserts where there are few earthworms.

2.2.6.4 Termitidae (Termites)

Termites have complex castes. Because they can consume wood, they have become economic pests in some locations. Termites make up 75% and 10% of tropical insect and terrestrial animal biomass, respectively. Kalotermitidae termites can digest cellulose thanks to their protozoan gut flora. They devour soil-contaminated wood. The Termitidae family can absorb and digest humified organic molecules in tropical soils without protozoan symbionts because of their wide spectrum of microbial symbionts.[17, 18]

2.3 INTERACTIONS AMONG THE BIOTIC COMMUNITY

Nutrient-rich soil is varied and complex. Microbes inhabit soil. Microorganisms decompose organic matter, recycle minerals, maintain soil structure, and more. These processes need interaction between soil micro- and macroflora, associative and antagonistic effects, and biotic and abiotic variables.

Heinrich Anton de Bary used the term *symbiosis* in 1878 to characterise species' beneficial or negative ecological relationships.[19] Environmental interactions are categorised by whether the symbiote benefits, harms or has no effect on the host:

1. "Positive ecological interactions" – commensalism, mutualism and symbiosis.
2. "Negative ecological interactions" – competition, amensalism, parasitism and predation. They may reach a material sooner or use it better genetically.
3. Neutralism – "neutral connection."

In this scenario, mutualism is a mutually advantageous contact between two partners, where at least one partner is mandatorily dependent on the other. Both partners can survive independently in the environment but can choose to profit from each other. Hawaii's *Myrica faya* (fayatree) is symbiotic with nitrogen-fixing soil. The actinomycete Frankia alters soil nitrogen in return for a place to live and carbon through its roots.

Commensalism occurs when symbionts use the host for their benefit without harming it. Bacteria compete for nutrients and energy. Mycorrhizal fungal infection may benefit the plant and supply carbon for fungi in low-phosphorus conditions. However, nitrogen limitation in plants may cause plant–microbe competition for nitrogen in soil.[20]

Amensalism is a negative ecological relationship between two creatures whereby one is destroyed while the other is unharmed. Its metabolites can also inhibit another microorganism[21] (Figure 2.1).

Parasitism, also known as the host–parasite connection, is an ecological interaction in which one organism or parasite uses resources to grow, reproduce

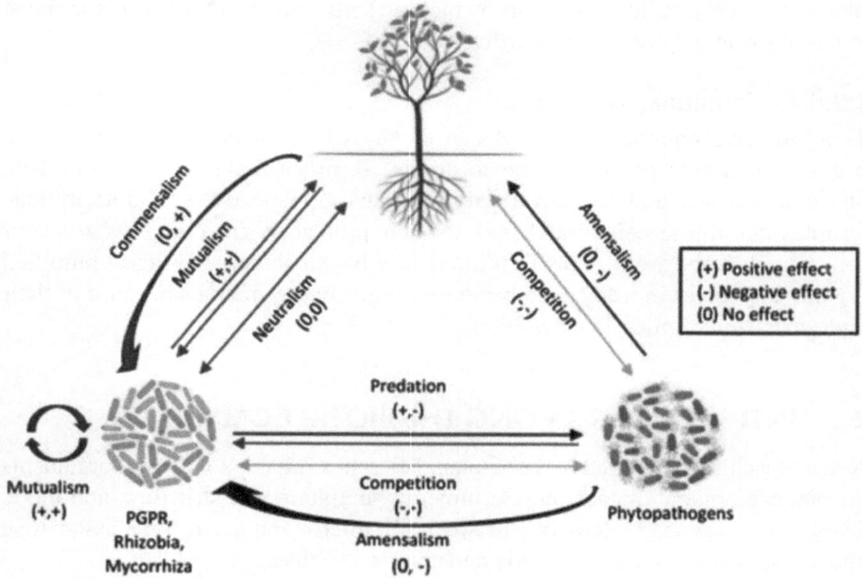

FIGURE 2.1 Rhizo-organisms and plant interaction (From Farooq, M., Wahid, A. and Siddique, K.H.M., *J. Soil Sci. Plant Nutr.*, 125, 12, 2012.)

and survive in the host cell while harming it. *Botrytis cinerea* (grey mould) and *Sclerotinia sclerotiorum* (stem rot) kill host cells while proliferating. The predator and prey exchange energy during predation. Predators kill animals for sustenance, yet decomposition, nutrient cycling, etc. can improve plant performance.[22]

2.4 SOIL FOOD WEB

Combining methods to examine soil biota in a food web setting can improve ecosystem functional prediction.[23] Better understanding soil food webs may help explain why soil biota dynamics vary so much.[24] Pathogens and mutualists can suppress or promote plant species, affecting the plant ecosystem. Microbes and soil fauna can alter the availability of nutrients to plants through their feeding activities.[25]

Ecology focuses on population and community dynamics. This is primarily connected to food interactions in biotic communities and the consequent forces that prey and predators apply to one another. Food webs depict biotic communities' complicated feeding connections. Food webs show prey–predator relationships by drawing arrows (Figure 2.2). These connection food webs are the most frequent, since they are simple to establish.[26] They show the key relationships between species but don't show how predators affect prey or vice versa, rendering them useless for understanding food web mechanics. Energy flow webs are a major development in food webs.

Energy flow webs have gained a lot of attention, notably in soil ecology. Community food webs link population dynamics and energetics, since both

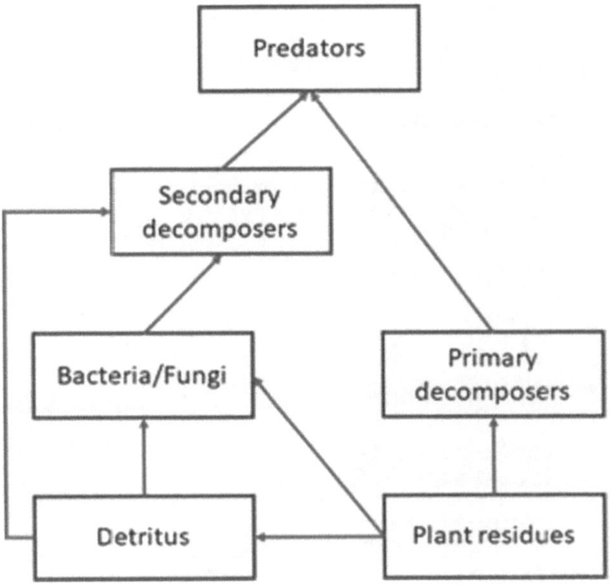

FIGURE 2.2 Below-ground food web chain of primary and secondary decomposers.

capture population interactions. These interactions alter population dynamics and reveal energy and matter transfer rates. The capacity to analyse community stability via modelling the interactions assemblage helps us determine whether all populations can continue and conserve biological diversity.

Primary producers – plants, lichens, mosses, photosynthetic bacteria and algae – create all food webs. Sun-powered animals repair atmospheric carbon dioxide.[27] Organic matter – humus and active organic material – is necessary for food web interactions, including breakdown. Most soil organisms get their energy and carbon from plants, animals and garbage. Chemoautotrophic bacteria get their energy from nitrogen, sulphur or iron instead of carbon or the sun. Many invertebrates control nutrition by eating organic substrates and materials. These materials break down faster when fragmented. This exposes cytoplasm and increases plant structural component surfaces, making them accessible. Consuming activity often transfers N in the form of excrement and fungal hyphae from soil to substrate, which accelerates breakdown. Many microorganisms attach to invertebrate exoskeletons, cuticles and digestive systems, transferring germs from one organic source to another.

Soil creatures inhabit several food chain trophic levels. Herbivores eat living plant roots, but most eat debris (saprophytes) and microbes. Some are carnivores. Soil food web analysis has examined species abundance, features and trophic resources. Trophic assessments are harder because soil food webs and above-ground systems are interconnected.[28] Even permanent residents may have adapted to soil depths. In Figure 2.3, "litter transformers," "ecosystem engineers" and "micro-food webs" build a soil food web.

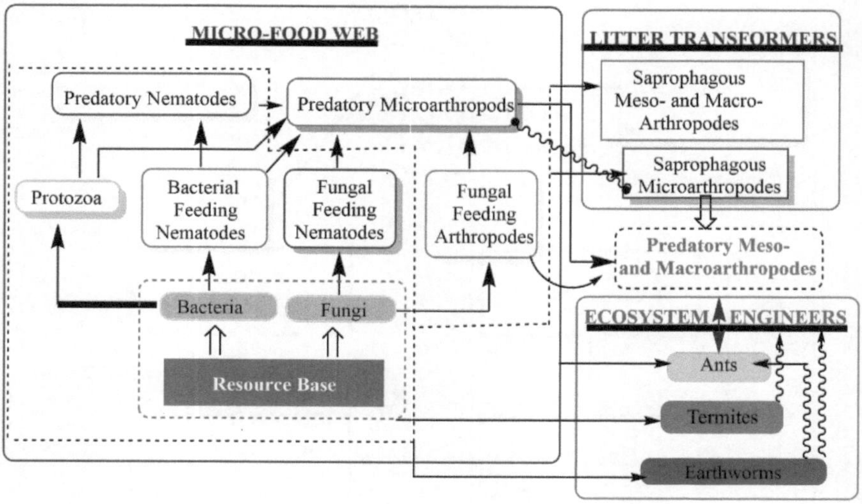

FIGURE 2.3 Soil food web organisation in three different categories.

2.5 SOIL ENVIRONMENT AND FACTORS AFFECTING SOIL BIOTIC COMMUNITY

Earth has skin – soil. Sustaining life on Earth depends on soil conditions. Since soil microbes are the foundation of soil ecological function, they help preserve ecological balance. Due to pH, salinity, temperature, humidity, soil water content and other conditions, soil crusts degrade faster. Soil organisms provide total organic carbon and nitrogen functions. C, N, P and S cycles define soil functioning. Soil bacteria decompose nutrients for other biotic components, maintaining the nutrient cycle. pH, moisture and salinity affect the soil biotic ecosystem and limit microorganisms. In wetland ecosystems, soil water content has a direct effect.[29]

Humus, which contains dead or living plants and animal leftovers, feeds microbes and contains easily digestible sugar and protein that help other biotic components thrive. Sand, silt and clay comprise soil. Physical, biological and chemical variables affect soil microorganism diversity and richness. The soil biotic community is affected by salinity, water content, pH and nutrients.

2.5.1 Soil nutrients

Soil nutrients, including carbon, nitrogen, phosphorus, sulphur and others, affect the biotic community's growth in the soil. Microorganism development decreases with soil nitrogen deficiency. Carbon adds structure and water retention to soil. Increasing the soil organic carbon store increases soil quality and sequesters more atmospheric carbon. Microbial biomass transfers nitrogen from microorganisms to plants and back to soil. The nitrogen cycle transforms it.[30]

2.5.2 pH

Soils have a pH of 3.5–10. Natural soil pH ranges from 5 to 7 in wet areas and from 6.5 to 9 in dry areas. Soil's biological, chemical and physical properties depend on pH. Soil pH affects precipitation, dissolution, mineralisation and more (Figure 2.4).

2.5.3 Salinity

Saline soil is saltier. Soil bacteria inhibit microorganisms and mineralise soil. The biotic community's osmotic potential is likewise affected. Salinity-tolerant bacteria synthesise osmolytes to regulate anabolism and catabolism and grow well. Na^+, Ca^+, Mg^+ and K^+ make up soil salts.

2.5.4 The water content of the soil

Soil holds much water based on two conditions. First, water content measures soil water; second, soil water potential measures how much energy the soil uses to hold water. Hydrolysis and transport require water. Soil moisture controls

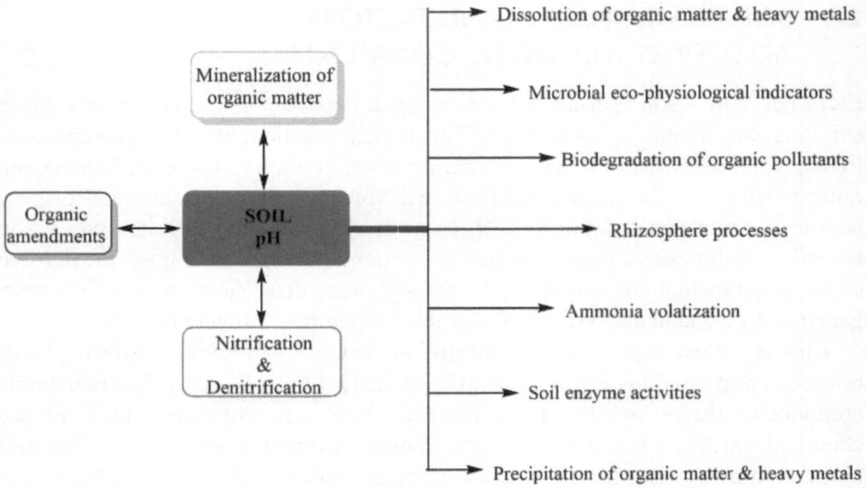

FIGURE 2.4 Different functions of soil pH.

microbial activity. Due to osmotic fluctuations, sensitive microbial cells may lose if soil water content varies.[31]

2.6 ROLE OF THE BIOTIC COMMUNITY OF SOIL

2.6.1 Role in the agriculture system

P and N availability limits agricultural output. Understanding how plant–microbe interactions affect nutrients will improve crop production's economic and environmental sustainability. Soil MOs affect P and N absorption in agriculture. Many soil microorganisms promote plant growth, and increasingly, commercial microbial products do too.[32] Plant growth-promoting microorganisms (PGPMs) penetrate roots, remain in the near-rhizoplane and stimulate plant growth. PGPMs increase biomass synthesis and disease resistance. PGPMs benefit from siderophores, phytohormones, antagonistic action and nutritional competition.[33] One PGPM, *Pantoea agglomerans*, stimulates maize growth but infects human soft tissue or bones/joints. *Burkholderia* can infect humans but also promotes plant growth, bioremediation and biological control.

2.6.2 Role in nitrogen fixation

N fixation changes soil biotic community and biogeochemistry. Soil N begins with root nodule organisms fixing N and integrating it into plant tissues. The N cycle in SAT habitats is affected by species like *Collema* sp. that create "biological soil crusts" (BSCs) and fix nitrogen. Biotic limitations on N cycling and transformations rarely consider BSCs.[34] Many ecosystems are shaped by nitrogen cycling. SAT habitats have a poor N cycle due to low N deposition and MO N fixation rates. Arid and semi-arid ecosystems provide 30% of global gaseous N emissions while having low N supply. BSCs – lichens, fungi, mosses and others – control the rest.

Soil microorganisms increase biofertilisation, plant root growth, rhizoremediation, abiotic stress reduction and infection control.[35] Fungi and bacteria produce lots of antibiotic compounds in the rhizosphere. Depth affects microbe–root connections. Most algae are on the surface; however, some blue-green algae are near organic-compound-producing roots. Many soil microorganisms protect plant roots after primary infection.[36, 37]

2.6.3 Role in mineralisation

Mineralisation converts biological nitrogen to NH_4^+ and NO_3. After decomposition, soil microorganisms release nutrients. They supply soil nutrients to plants.[38] Organic matter with a C to N ratio of <22:1 increases soil mineral nitrogen. When introducing organic matter, microbes often extract mineral nitrogen from the soil at C:N ratios >22:1. Soil bacteria provide N, S and P to plant roots.[39]

Glycerine, cysteine and carbon dioxide are released when microbes break down plant proteins. NH_4^+, NO_3 and SO_4^{2-} are produced by further degrading these amino acids' nitrogen and sulphur. The soil's food source depletes microorganisms. Soil bacteria eat dead microorganisms. Water, mineral nutrients (nitrogen, phosphorus, sulphur and trace elements) and carbon dioxide also enter the soil.[40]

2.6.4 Role in nitrification and denitrification

Nitrification is the enzymatic conversion of ammonia to nitrates by soil microorganisms. Unless microorganisms convert nitrite to nitrate, plants and mammals are at risk. Nitrate-producing microorganisms are sensitive. Nitrification requires oxidised ammonia. High C/N residue ratios also inhibit nitrification. Soil aeration and drainage must give oxygen for nitrification. Low and high soil moisture slow nitrification. Low pH soil hinders nitrification, which depends on cations that form bases. Soil pesticides inhibit nitrification. Negatively charged soil colloids do not absorb nitrate ions. They evaporate from the soil's deeper horizons.

Facultative anaerobic microbes reduce nitrate nitrogen to gaseous molecules. Reductase enzymes aid denitrification. NO, N_2O and N_2 can be released. Anaerobic bacteria absorb oxygen. Soil nitrogen can be molecularised by non-microbial mechanisms. Wet soil denitrification losses and those induced by poor aeration may be particularly significant if soil nitrogen levels may be decreased.

2.6.5 Role in sulphur immobilisation

Microorganisms immobilise sulphur in soils from low-S organic materials. It works like nitrogen mineralisation. Energy-dense organic residues stimulate microbial development and create sulphate. After death, soil sulphate becomes inorganic. Hydrogen sulphide, carbon disulphide, carbonyl sulphide and mercaptan are formed when microbes decompose organic compounds. Anaerobic soils are high in all of these. Most are made by bacteria breaking down S-containing amino acids like methionine and cysteine. Some gases are absorbed by soil colloids, whereas others escape and return after chemical changes.[41]

2.7 CONCLUSION

The soil biotic community is significantly influenced by various biotic factors that limit the growth of microbes. The present chapter gives a clear understanding of soil as a host to the biotic community, and it may assist researchers in clarifying why there are such significant differences in soil-type-specific soil biota dynamics.

ACKNOWLEDGEMENT

The authors would like to thank Dr. Mital D. Jasani, WOSA, ICAR-DGR, Junagadh, Gujarat, India for insightful discussions and providing assistance in the MS preparation.

REFERENCES

1. Yang, Gaowen, Cameron Wagg, D. Stavros, et al. "How soil biota drive ecosystem stability." *Trends in Plant Science* 23, no. 12 (2018): 1057–1067.
2. Manuel, Delgado-Baquerizo, Angela M. Oliverio, Tess E. Brewer, et al. "A global atlas of the dominant bacteria found in soil." *Science* 359, no. 6373 (2018): 320–325.
3. Hayat, Rifat, Safdar Ali, Ummay Amara, et al. "Soil beneficial bacteria and their role in plant growth promotion: A review." *Annals of Microbiology* 60, no. 4 (2010): 579–598.
4. Guo, Junjie, Ning Ling, Zhaojie Chen, et al. "Soil fungal assemblage complexity is dependent on soil fertility and dominated by deterministic processes." *New Phytologist* 226, no. 1 (2020): 232–243.
5. Bridge, Paul, and Brian Spooner. "Soil fungi: Diversity and detection." *Plant and Soil* 232, no. 1 (2001): 147–154.
6. Finlay, Roger D., and R. Greg Thorn. "The fungi in soil." *Modern Soil Microbiology* (2019): 65–90.
7. Coleman, David C., and Diana H. Wall. "Soil fauna: Occurrence, biodiversity, and roles in ecosystem function." *Soil Microbiology, Ecology and Biochemistry* 4 (2015): 111–149.
8. Haynes, Richard John. "Nature of the belowground ecosystem and its development during pedogenesis." *Advances in Agronomy* 127 (2014): 43–109.
9. Esteban, Genoveva F., and Tom M. Fenchel. "Protozoan communities: Terrestrial habitats." In *Ecology of Protozoa*, pp. 157–174. Springer, Cham, 2020.
10. Robeson, Michael S., J. Andrew, Kristen R. Freeman, et al. "Soil rotifer communities are extremely diverse globally but spatially autocorrelated locally." *Proceedings of the National Academy of Sciences* 108, no. 11 (2011): 4406–4410.

11. Dervash, M., R. Bhat, Nighat Mushtaq, and V. Singh. "Dynamics and importance of soil mesofauna." *International Journal of Advance Research in Science and Engineering* 7, no. 4 (2018).
12. Wall, Diana H., Valerie Behan-Pelletier, T. Hefin Jones, et al. van der Putten. In *Soil Ecology and Ecosystem Services*. Oxford University Press, 2012.
13. Lakshmi, Gopakumar, Bernard N. Okafor, and Donato Visconti. "Soil microarthropods and nutrient cycling." In *Environment, Climate, Plant and Vegetation Growth*, pp. 453–472. Springer, Cham, 2020.
14. Jeffery, S., C. Gardi, A. Jones, L. Montanarella, L. Marmo, L. Miko, K. Ritz, G. Peres, J. Römbke, and H. R. S. G. Van der Putten. *European atlas of Soil Biodiversity. European Commission*. Publications of the European Union, Luxemburg, 2010.
15. Pelosi, C., P. Thiel, S. Bart, et al. "The contributions of enchytraeids and earthworms to the soil mineralization process in soils with fungicide." *Ecotoxicology* 30, no. 9 (2021): 1910–1921.
16. Bagyaraj, D. J., C. J. Nethravathi, and K. S. Nitin. "Soil biodiversity and arthropods: Role in soil fertility." In *Economic and Ecological Significance of Arthropods in Diversified Ecosystems*, pp. 17–51. Springer, Singapore, 2016.
17. Culliney, Thomas W. "Role of arthropods in maintaining soil fertility." *Agriculture* 3, no. 4 (2013): 629–659.
18. Palin, Olivia F., Paul Eggleton, Yadvinder Malhi, et al. "Termite diversity along an Amazon–Andes elevation gradient, Peru." *Biotropica* 43, no. 1 (2011): 100–107.
19. Bonfante, Paola, K. Visick, and M. Ohkuma. "Symbiosis in the omics era." *Environ. Microbiol. Rep.* 2 (2010): 475–478.
20. Smith, Sally E., and David J. Read. *Mycorrhizal Symbiosis*. Academic Press, 2010.
21. Barbato, Robyn A. "The soil habitat and considerations for synthetic biology." In *Emerging Threats of Synthetic Biology and Biotechnology*, pp. 169–175. Springer, 2021.
22. Bonkowski, Michael, and Marianne Clarholm. "Stimulation of plant growth through interactions of bacteria and protozoa: Testing the auxiliary microbial loop hypothesis." *Acta Protozoologica* 51, no. 3 (2012).
23. Jamil, Fatima, Hamid Mukhtar, Mireille Fouillaud, and Laurent Dufossé. "Rhizosphere signaling: Insights into plant–rhizomicrobiome interactions for sustainable agronomy." *Microorganisms* 10, no. 5 (2022): 899.
24. Birkhofer, Klaus, Ingo Schöning, Fabian Alt, et al. "General relationships between abiotic soil properties and soil biota across spatial scales and different land-use types." (2012): e43292.
25. Maron, John L., Marilyn Marler, John N. Klironomos, and Cory C. Cleveland. "Soil fungal pathogens and the relationship between plant diversity and productivity." *Ecology Letters* 14, no. 1 (2011): 36–41.
26. Morriën, E. "Understanding soil food web dynamics, how close do we get?" *Soil Biology and Biochemistry* 102 (2016): 10–13.
27. Khatoon, Hina, Praveen Solanki, Maitrayie Narayan, et al. "Role of microbes in organic carbon decomposition and maintenance of soil ecosystem." *International Journal of Chemical Studies* 5, no. 6 (2017): 1648–1656.
28. Moore, John C., and Peter C. De Ruiter. "Soil food webs in agricultural ecosystems." *Microbial Ecology of Sustainable Agroecosystems*, pp. 63–88. CRC Press, 2012.
29. Yan, Nan, Petra Marschner, Wenhong Cao, Changqing Zuo, and Wei Qin. "Influence of salinity and water content on soil microorganisms." *International Soil and Water Conservation Research* 3, no. 4 (2015): 316–323.
30. Neina, Dora. "The role of soil pH in plant nutrition and soil remediation." *Applied and Environmental Soil Science* 2019 (2019).

31. Simpson, Jake E., Eleanor Slade, Terhi Riutta, and Michele E. Taylor. "Factors affecting soil fauna feeding activity in a fragmented lowland temperate deciduous woodland." *PLoS One* 7, no. 1 (2012): e29616.
32. Antunes, Pedro M., Philipp Franken, Dietmar Schwarz, et al. "Linking soil biodiversity and human health: Do arbuscular mycorrhizal fungi contribute to food nutrition." *Soil Ecology and Ecosystem Services* (2012): 153–172.
33. Montañez, Adriana, Andrea Rodríguez Blanco, Claudia Barlocco, Martin Beracochea, and Margarita Sicardi. "Characterization of cultivable putative endophytic plant growth promoting bacteria associated with maize cultivars (Zea mays L.) and their inoculation effects in vitro." *Applied Soil Ecology* 58 (2012): 21–28.
34. Castillo-Monroy, Andrea P., T. Fernando, Manuel Delgado-Baquerizo, et al. "Biological soil crusts modulate nitrogen availability in semi-arid ecosystems: Insights from a Mediterranean grassland." *Plant and Soil* 333, no. 1 (2010): 21–34.
35. Qiang, Xiaoyu, Michael Weiss, Karl-Heinz Kogel, and Patrick Schäfer. "Piriformospora indica: A mutualistic basidiomycete with an exceptionally large plant host range." *Molecular Plant Pathology* 13, no. 5 (2012): 508–518.
36. Mela, Francesca, Kathrin Fritsche, Wietse De Boer, et al. "Dual transcriptional profiling of a bacterial/fungal confrontation: Collimonas fungivorans versus Aspergillus niger." *The ISME Journal* 5, no. 9 (2011): 1494–1504.
37. Pieterse, Corné M. J. "Prime time for transgenerational defense." *Plant Physiology* 158, no. 2 (2012): 545–545.
38. Farooq, Muhammad, A. Wahid, and Kadambot HM Siddique. "Micronutrient application through seed treatments: A review." *Journal of Soil Science and Plant Nutrition* 12, no. 1 (2012): 125–142.
39. Hoyle, Francis C., Jeff A. Baldock, and Daniel V. Murphy. "Soil organic carbon–role in rainfed farming systems." In *Rainfed Farming Systems*, pp. 339–361. Springer, 2011.
40. Pandey, Shyam Narain, Murtaza Abid, and Mirza Mohammad Abid Ali Khan. "Diversity, functions, and stress responses of soil microorganisms." *Plant Microbiome: Stress Response* (2018): 1–19.
41. Zhou, Xiaoqi, Zhiying Guo, Chengrong Chen, and Zhongjun Jia. "Soil microbial community structure and diversity are largely influenced by soil pH and nutrient quality in 78-year-old tree plantations." *Biogeosciences* 14, no. 8 (2017): 2101–2111.

3 Effects of soil characteristics on the diversity and biomass of microbial communities

Mohd. Shahnawaz[#], Manisha K. Sangale[#], Beenish Sadaqat[#], Mudasir Ahmad Dar[*], Sehrish Gazal, Romana Akhtar, Mohd. Yaseen Sirwal, Puja Gupta[*] Rongrong Xie and Jianzhong Sun

CONTENTS

3.1 INTRODUCTION

Soil is a crucial component of all terrestrial ecosystems.[1] Soil is an interface of the lithosphere, hydrosphere and atmosphere that supports life. Soil is known to be one of the most versatile ecosystems on planet Earth, consisting of flora, fauna and varied microbial taxa, which contribute to soil formation and productivity by carrying out innumerable processes. Soil microbes are the largest and

[*] Corresponding author: muddar7@gmail.com; pujapranavmahajan@gmail.com
[#] Authors contributed equally and are all first authors

DOI: 10.1201/9781003354031-3

most diversified biotic group in soil, having an estimated 1 million to 10 billion microbes of various species in 1 g of agricultural soil.[2, 3] For the proper balance and maintenance of life, soil quality is imperative for sustainability. Soil quality is usually described as the capability of soil to sustain biological productivity, preserve environmental quality, and promote plant and animal health and growth. Compared with the quality of air and water, soil quality is more complex due to the complexity of the soil ecosystems, involving various biotic and abiotic features.[4] In soil ecosystems, living organisms play a central role in their functioning. It has been presented that several biotic and abiotic characteristics, such as soil pH, Mg, K, total N content, microbial biomass, and their community structure and functions, contribute to the overall properties of soils.[5] Among living organisms, microbes, including bacteria and fungi, are found in every environment.[6] Regarding the stability of the ecosystem, the most essential part of the terrestrial ecosystem may be the microbial biomass. The pH, soil moisture, organic matter content, floral composition and climatic conditions all affect the soil's microbial diversity. The bacteria, fungi and other microfauna that make up the soil's microbial population are termed the *soil microbial biomass* (SMB). This pool of organisms in the soil is measured as the amount of C and N and accordingly represented as SMB-C and SMB-N. The SMB ranges from 110 kg-C/ha in a cultivated field to 2240 kg-C/ha in a grassland soil with total C of 7.0%.[7] The present chapter critically reviews the soil quality and its microbial biomass. Further, the effect of soil texture, organic matter and other factors on microbial biomass is detailed. Lastly, we focus on the microbial biomass and its correlation and interactions with soil quality.

3.2 MICROBIAL BIOMASS AND SOIL FUNCTION

Microbes help in the formation of granular structure of soil by binding soil particles with organic matter.[8] Amongst the diverse microbial biota, such as bacteria, protozoa, fungi and actinomyces, bacteria are found to be the smallest and most varied. Bacteria are more prevalent in grasslands and agricultural soils; however, fungi prefer acidic soils and forest areas.[9] Soil microorganisms perform a key role in the decomposition of organic matter and recycling of nutrients, such as nitrification, ammonification, oxidation and nitrogen fixation.[10] They also have been shown to boost soil aggregate stability and improve plant water relations. As decomposers, symbionts and pathogens, fungi comprise a key component of soils and play a significant role in the diversity and functional restoration of ecosystems.[11] Compared with bacteria, fungal communities are more sensitive to non-extreme fluctuations in soil moisture. When soil moisture varies, the structure of fungal communities also varies. Hitherto, the effects of severe drought and heavy rain events on the variety and functioning of soil have been the subject of several research efforts, but less attention has been given to how microbial communities react to less extreme fluctuations in soil moisture.[12] Fungi in forest soil communities exhibit a wide range of different lifestyles, from saprophytes to endophytes, pathogens and mycorrhizal species. They are incredibly complex and diverse. These underground communities are involved in important ecosystem processes like plant community dynamics, below-ground trophic interactions and biogeochemical cycles. They are distributed differently across space and time scales. Using instances from several global forest ecosystems, we outline the key biotic and abiotic elements, such as natural disturbances, that influence the distribution of fungal communities in forest soils as well as potential underlying mechanisms.[13] Fungi can be found practically anywhere and are crucial to ecosystem processes including nitrogen recycling and degradation. They play a significant role in maintaining biodiversity on a global scale and regulate the speed of important ecosystem processes, with fungi predominating in the breakdown of lignified cellulose and other plant materials. They produce a wide variety of extracellular enzymes that disassemble complex organic polymers into simpler shapes so that the fungi or other organisms can absorb them.[14] Therefore, it is not surprising that removing fungi leads to a considerable reduction in both carbon and nitrogen depletion from litter. In fact, fungal hyphae often account for the greatest fraction of soil biomass and can reach hundreds to thousands of milligrams per gram of soil. Through a variety of interactions, fungi directly affect plants, animals and microorganisms. They are the most prevalent and significant plant disease agents, causing severe crop losses, and alter the content and structure of natural plant communities.[15] Despite their importance, the majority of fungi have not been identified yet. In addition to being dynamic, fungal communities are biologically diverse, containing both species of free-living saprobes and mycorrhizal mutualists that are connected with plants. The makeup of the mycorrhizal community may have changed due to species-specific ability to maintain an adequate nutrient supply to the host, and these changes may thus have functional components.[16] Saprotrophic fungi's composition and population size can be significantly influenced by plant diversity, offering a huge range of substrates to create helpful interactions with them. However, as fungal species are filtered with shifting environmental conditions, global climate change has the potential to drastically disrupt these communities.[17] Particularly in temperate zones, the two main fungal guilds in forest ecosystems are ectomycorrhizal (ECM) and saprophytic fungi. Through mutualistic interactions with plant roots, ECM fungi determine how well nutrients and water are absorbed by plants. The primary agents in the breakdown of complex organic matter and nutrient conversion are saprophytic fungi. The composition of saprotrophic fungi can be significantly influenced by plant diversity, which can also boost the diversity of their communities by offering a wide range of substrates for establishing helpful interactions with them.[18] Microorganisms in the soil are crucial to the functioning of the soil ecosystem. Importantly, soil microbial diversity and organisation have been recognised as key determinants of soil quality and functionality. Fungi play a crucial role in the soil that impacts numerous ecological processes, such as the cycling of nutrients and vital components including carbon, nitrogen, phosphorus and sulphur. In turn, these factors also demonstrate a vital role in shaping the fungal community of the soils. Some of the factors that affect the SMB are discussed in detail in the following sections.[19]

3.3 EFFECT OF SOIL PROPERTIES ON MICROBIAL BIOMASS

Physicochemical and biological properties of soil, like parent material, texture, temperature, oxygen levels, pH, moisture, organic matter and clay content, determine the diversity and biomass of soil biota (Figure 3.1) by providing nutritional requirements for the development of the microbial community.[20] The different soil horizons have different microbial populations, which are responsible for numerous ecosystem functions. However, little is known about how soil horizons affect microbial communities. The change in the abundance and diversity of the microbial communities in the soil is dependent on the depth as well as other features of the soil. Recently, Luo et al.[21] stated that the species richness and diversity of the soil fungi varied with soil horizons, being higher in the O horizons. Concentrations of P, K, Ca, Mg, pH and nitrate in the soil not only vary with the soil depth but also cause noticeable impacts on the soil microbiota.[22, 23] Thus, any change in the soil's biotic and abiotic characteristics may directly affect the soil microbiota.[24, 25]

3.3.1 Effect of soil depth on microbial biomass

The soil profile primarily exhibits a significant amount of diversity in its structure and composition (Table 3.1). The soil vertical profile acts as an ecological filter for the distribution of microorganisms by rendering highly heterogeneous environments and exhibiting varied soil properties, hence affecting microbial diversity, composition and functions.[26] Soil characteristics like moisture, texture, porosity and bulk density tend to vary with the soil depth. Soils get compacted and exhibit greater clay content with the increase in depth, causing subsoils to exhibit higher bulk density and low porosity in comparison to the surface soils.[27] This results in low gas diffusion and root growth, thereby making preferable microclimates for the growth of anaerobic and non-root-associated microbiota[28] (Figure 3.2). The distribution and composition of soil microbes decrease substantially with depth, mainly due to greater availability of carbon content in the upper layers of the soil than in the lower.[29]

Around 65% of the total microbial biomass is observed down to the depth of 25 cm, below which the microbial biomass tends to decrease greatly.[30, 32] Gram-negative bacteria, fungi and protozoa usually are more abundant on the surface layers in comparison to Gram-positive bacteria, actinobacteria, and archaea, which are found more abundantly in the deeper soils.[31] The distribution of Gram-negative and Gram-positive bacteria is greatly linked to their nutritional preferences. The oligotrophic Gram-positive bacteria dominate subsoils that are nutrient deficient as compared with surface layers that are more abundant in carbon sources and are dominated by Gram-negative bacteria.[33]

Therefore, soil microfloral and microfaunal populations are governed mainly by soil organic matter, depth and aeration. It has been discovered that the microbial population thrives better in surface soils that are well-aerated and rich in organic content, favouring the growth of microbes (Table 3.2). Additionally, as

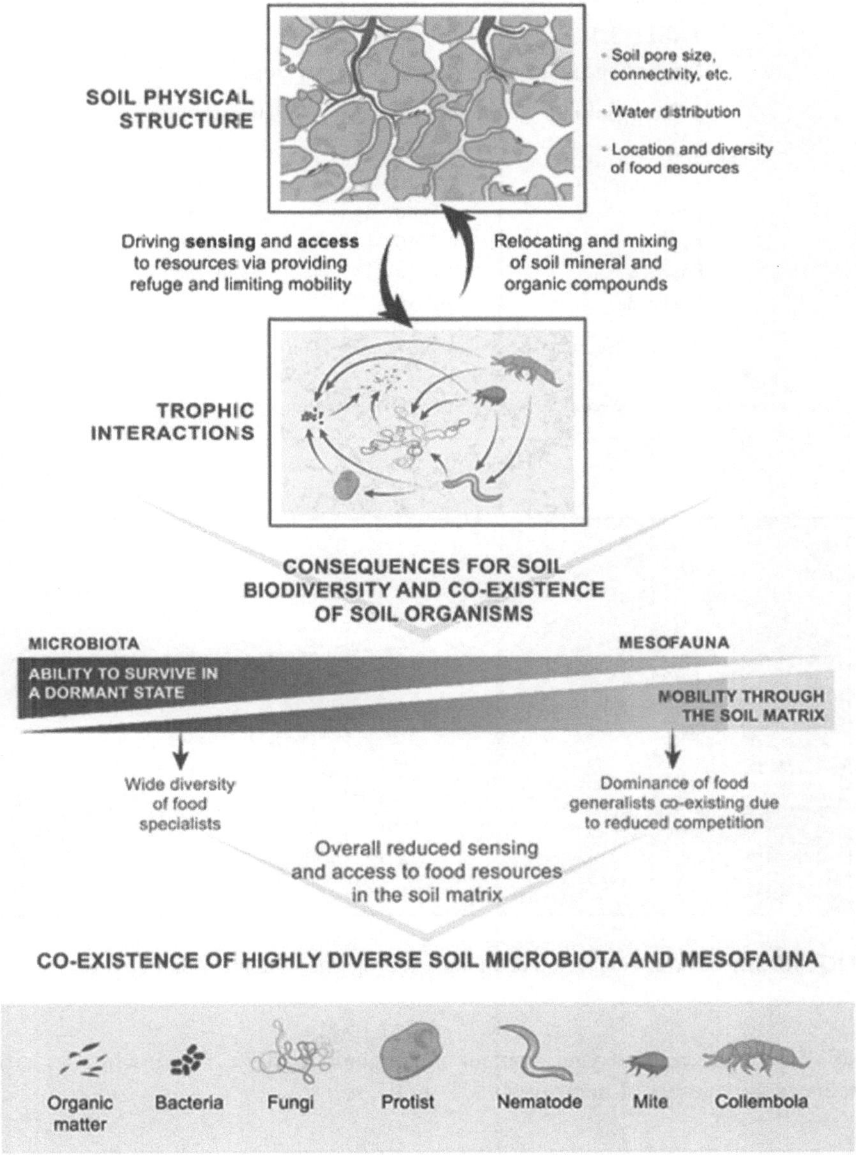

FIGURE 3.1 Overview of the effects of soil physical structure on trophic interactions and soil processes.

TABLE 3.1
Microorganism species at 0–15 cm depth of soil

Microorganism	Number/g soil	Biomass (g/m²)
Bacteria	10^8–10^9	40–500
Actinomycetes	10^7–10^8	40--500
Fungi	10^5–10^6	100–1500
Algae	10^4–10^5	Jan-50
Protozoa	10^3–10^4	Varies
Nematodes	10^2–10^3	Varies

Source: Hoorman, J.J., *Understanding Soil Microbes and Nutrient Recycling: SAG-16, Agriculture and Natural Resources.*, Ohio State University, Ohio, 2010.

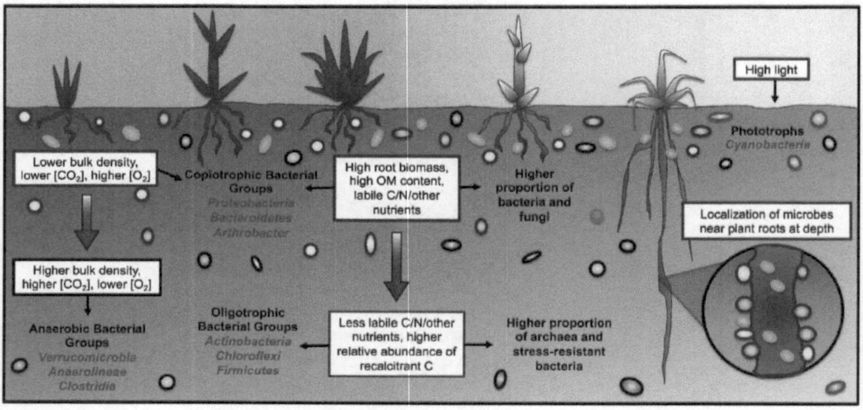

FIGURE 3.2 The effects of soil depth on soil properties and microbial groups.

soil depth increases, oxygen content and organic matter decline, which in turn reduces the number of microorganisms in the soil.[34]

3.3.2 Effect of soil texture on microbial biomass

The texture of the soil has a great impact on the availability and retention potential of different nutrients in the soil and also plays a key role in carbon sequestration. Soil type and structure have a pronounced effect on soil microbial population by modulating the availability of nutrients for carrying out various biological processes. Clay-loam and silty-clay-loam soils are known to support larger and more diverse bacterial communities as compared with sandy-loam and silty soils,

TABLE 3.2
Decreasing microorganism population with increasing soil depth

Profile	Soil Depth (cm)	Microorganism Biomass (g/m^2)
Terrace	0 to 5	9.8 ± 1.6
	5 to 15	4.0 ± 0.16
	15 to 25	2.0 ± 0.12
	50	0.63 ± 0.044
	100	0.18 ± 0.030
	200	0.081 ± 0.0053
Valley	0 to 5	16 ± 0.040
	5 to 15	5.1 ± 0.41
	15 to 25	2.5 ± 0.16
	50	0.84 ± 0.077
	100	0.41 ± 0.093
	200	0.11 ± 0.043

Source: Fierer, Noah, Joshua P. Schimel, and Patricia A. Holden. *Soil Biology and Biochemistry* 35, no. 1 (2003): 167–176.

which are mostly dominated by fungi.[35] Soils with greater clay content have more macro-aggregates due to the presence of a larger number of reactive surfaces, which aids in the formation of organo-mineral complexes. Greater clay content in the soil also provides better microbial biomass protection, holds a substantial amount of water[36] and tends to show a positive relationship with the nutrient retention potential.[37] Clay particles also tend to protect soil microbiota against predation, dehydration, heat and pH alterations. Higher species richness of bacterial communities has been seen in coarsely structured soils due to the presence of large pore size and a higher number of water films.[38] As pores become smaller and less frequent with increasing soil depth, this leads to restricted movement of microbes, nutrients and oxygen, thereby resulting in the heterogeneous and confined activity of microorganisms in specific regions having greater inputs of moisture and substrate.[39]

The soil texture can regulate nutritional and environmental stress encountered by soil microbial communities and govern the establishment of the microbiota.[40] The smaller pore size and slow mobility of bacteria within the soils having small pore size results in a higher diversity of bacterial communities due to increased segregation. Fungi, however, have the potential to move from dry pore spaces of soil, and hence, their movement is not restricted to a particular soil type.[41]

3.3.3 Effect of Organic Matter on Microbial Population

Soil organic matter is a critical component of soil health because of the numerous fundamental services it offers and supports.[42] Hence, soil organic matter is considered as an indicator of ecosystem functioning. Reduction in soil organic matter leads to a decrease in biodiversity and soil fertility, increased soil erosion and soil compaction.[43] The majority of bacteria, fungi and actinomycetes are found to correlate positively with soil's organic carbon content. Carbon sources provide energy for microbes, thus making carbon a limiting factor for the establishment of the microbial communities. The distribution of soil microbiota is directly linked to the concentration of organic matter in the soil. Soils with higher organic matter are found to have higher microbial biomass and functional diversity.[44] Many studies have shown that soil microbial community structure and diversity are related to the variations in the soil organic carbon,[45] and the microbial biomass is known to decrease with soil depth due to the decrease in the concentration of organic matter. It is generally known that fungi thrive well on organic material having more complicated structural characteristics; however, bacteria flourish better in accessible organic carbon.[46]

3.3.4 Soil pH

Soil pH is a very important factor in determining the physical, chemical and biological properties of soils.[47] Numerous studies have determined the effect of pH on soil microbial communities, diversity and biomass, which demonstrated that soil pH strongly affects all these soil properties.[48] To say the least, pH is as important as soil C and N availability in determining the amount of SMB. Soil pH is also very significant for the solubility of soil C;[49] besides, with decreasing pH, biologically toxic Al concentrations increase in the soil environment.[50] The labile C and Al concentrations in turn affect the soil microbial activity and diversity.[51] Generally, the highest microbial diversity is found in forest soils with neutral pH compared to more acidic and basic pH. Fierer and Jackson[52] collected 98 soil samples from different regions of North and South America and demonstrated that soil pH greatly influences soil microbial structure. The soils with neutral pH possessed the highest microbial diversity, whereas acidic soil had the lowest number of microbes.

3.3.5 Soil moisture

Soil moisture is another factor that affects microbial biomass. Adequate moisture levels are important for maintaining soil homeostasis and also for maintaining soil microbial enzymes in an active state.[53] A moistened soil holds adequate microbial biomass up to a certain threshold. Beyond this threshold, increased moisture concentrations can adversely affect aerobic microbes.[54] The decreased microbial biomass in soils holding water levels may be due to depletion of oxygen, which creates unfavourable conditions for both bacteria and fungi.[55] However, under certain conditions, anaerobic microorganisms may increase in number. Naturally, the soil water profile fluctuates according to seasonal changes, rainfall, irrigation and temperature. In particular, semi-arid and Mediterranean soils face more frequent water fluctuations with long dry spells, which are followed by swift wet spells.[56] Fierer and Schimel[57] studied the effect of changing moisture conditions on soils and demonstrated that the activity of microbes peaked after 24 hours of rewatering. This may be because sensitive microbial cells lyse upon soil rewatering, whereas other soil microbes release organic molecules that were accumulated inside the cells during the dry phase.[58] Moreover, more organic matter is released into the soil ecosystem when soil aggregates break down; this can be decomposed by the available microbes in the soil, which in turn increases their population in the rewatered soil.[58] However, an increased number of drying and rewetting cycles can result in decreased biomass as a result of loss of C and higher biomass turnover number.[57] Moreover, these fluctuating conditions can also result in loss of some microbial communities and change in soil microbiota, which in turn can affect nutrient cycling in the soil. However, the response to changing moisture level and number of drying and rewatering spells varies with the type of soil.[59]

3.3.6 Nitrogen and phosphorus

The anthropogenic use of nitrogen has greatly increased since the beginning of the industrial revolution.[60] The increased amount of N in the soil may help to cope with nitrogen limitation and therefore increase the microbial biomass, in particular nitrifying and denitrifying bacteria, in the soil; however, an excess of N can cause negative effects on the soil environment in general and microbes in particular.[61] Excess N can be toxic to microbes that are sensitive to osmotic potential, thus causing decreased biomass.[62] The soil nitrogen availability also affects soil C availability, pH and plant diversity, which in turn influences soil microbiota. For example, the increased amount of soil nitrogen could enhance soil C availability,[63] which could help to increase the ratio of microbes in the soil that were previously C limited. On the other hand, increased N can also enhance the number of fine roots, which could increase the amount of lignin-like recalcitrant compounds in the soil, therefore decreasing the amount of available C in the soil. Furthermore, N addition has been found to decrease soil pH,[64] enhancing the leaching or mobilisation of magnesium, calcium and aluminium. Aluminium poisoning besides magnesium and calcium limitation in a nitrogen environment may decrease the diversity of soil microbes.[65] In this regard, microbes capable of utilising complex compounds may increase in number, whereas other microbial communities may either decline or disappear.

Anthropogenic activities have drastically increased the use of phosphorus in the soil. Compared with the pre-industrial era, P concentrations have increased dramatically nowadays. It is important to note that a quick increase of N and P may significantly affect several processes and fluxes taking place in the soil ecosystem.[66] In particular, the increased N and P concentrations drastically affect soil C availability and respiration. There are numerous studies available in the literature that have examined how P concentrations affect associated microbial communities in different soil ecosystems like grasslands, Arctic tundra, agricultural lands, forest soil and greenhouse areas.[67] Moreover, changes in SMB have also been examined relative to the type of fertiliser used to enhance the soil P concentrations, including mineral, rock, organic, and a mixture of organic and mineral fertilisers.[68] Moreover, the change in the microbial communities with changing P concentrations is strongly influenced by the host plants; for example, increased P concentrations strongly influenced microbial communities' composition in tropical forests.[69]

3.4 CONCLUSIONS

The chemical composition and microbial biomass are strongly impacted by changes in the soil profiles due to various factors based on the environment and eco-geographical locations. The foundation for comparing soils is based on the idea that soil moisture, pH, total carbon, nitrogen and phosphorus content, mineral content and organic matter have an effect on microbial population, soil depth and soil texture, which affect the microbial distribution in the soil that underlies ecosystem services. A promising attempt was made to correlate particular soil types, soil properties, soil restrictions and soil attributes to ecosystem services.

ACKNOWLEDGEMENT

MS and MKS are thankful to University of Ladakh, Ladakh, India and Natural Resource Wales, Swansea, United Kingdom for financial assistance. BS acknowledges Shandong University of Technology, China for postdoctoral fund. MAD acknowledges National Natural science foundation of China (3221101169). PG is thankful to DST, India for Women Scientist-A award.

REFERENCES

1. Young, I. M., and J. W. Crawford. Interactions and self-organization in the soil-microbe complex. *Science* 304, no. 5677 (2004): 1634–1637.
2. Wani, Fozia Shafiq, Latief Ahmad, Tahir Ali, and Andleeb Mushtaq. "Role of microorganisms in nutrient mobilization and soil health: A review." *Journal of Pure and Applied Microbiology* 9 (2015): 1401–1410.
3. Rosselló-Mora, Ramon, and Rudolf Amann. "The species concept for prokaryotes." *FEMS Microbiology Reviews* 25, no. 1 (2001): 39–67.
4. Pham, T. G., Nguyen, H. T., and M. Kappas. "Assessment of soil quality indicators under different agricultural land uses and topographic aspects in Central Vietnam." *International Soil and Water Conservation Research*, 6, no. 4 (2018): 280–288.
5. Wu, Guoju, Xiaohong Liu, Tuo Chen, Guobao Xu, Wenzhi Wang, Xiaomin Zeng, and Xuanwen Zhang. "Elevation-dependent variations of tree growth and intrinsic water-use efficiency in Schrenk spruce (Picea schrenkiana) in the western Tianshan Mountains, China." *Frontiers in Plant Science* 6 (2015): 309.
6. Kumar, Kalpna V., Shubhi Srivastava, N. Singh, and H. M. Behl. "Role of metal resistant plant growth promoting bacteria in ameliorating fly ash to the growth of Brassica juncea." *Journal of Hazardous Materials* 170, no. 1 (2009): 51–57.
7. Gatiboni, Luciano Colpo, Jefferson Luis Meirelles Coimbra, Rosiane Berenice Nicoloso Denardin, and Leandro do Prado Wildner. "Microbial biomass and

soil fauna during the decomposition of cover crops in no-tillage system." *Revista Brasileira de Ciência do Solo* 35 (2011): 1051–1057.

8. Six, J. A. E. T., Edward T. Elliott, and Keith Paustian. "Soil macroaggregate turnover and microaggregate formation: A mechanism for C sequestration under no-tillage agriculture." *Soil Biology and Biochemistry* 32, no. 14 (2000): 2099–2103.
9. Tugel, A. J., and A. M. Lewandowski. *Soil Biology Primer.* NRCS Soil Quality Institute, Ames, 1999.
10. Vymazal, Jan. "Removal of nutrients in various types of constructed wetlands." *Science of the Total Environment* 380, no. 1–3 (2007): 48–65.
11. Wardle, David A., Richard D. Bardgett, et al. "Ecological linkages between aboveground and belowground biota." *Science* 304, no. 5677 (2004): 1629–163.
12. Umair, Muhammad, Ningxiao Sun, Hongmei Du, et al. "Bacterial communities are more sensitive to water addition than fungal communities due to higher soil K and Na in a degraded karst ecosystem of Southwestern China." *Frontiers in Microbiology* 11 (2020): 562546.
13. Fierer, Noah, Joshua P. Schimel, and Patricia A. Holden. "Variations in microbial community composition through two soil depth profiles." *Soil Biology and Biochemistry* 35, no. 1 (2003): 167–176.
14. Frąc, Magdalena, Silja E. Hannula, Marta Bełka, and Małgorzata Jędryczka. "Fungal biodiversity and their role in soil health." *Frontiers in Microbiology* 9 (2018): 707.
15. Treseder, Kathleen K., and Michael F. Allen. "Mycorrhizal fungi have a potential role in soil carbon storage under elevated CO_2 and nitrogen deposition." *The New Phytologist* 147, no. 1 (2000): 189–200.
16. Rillig, Matthias C., and Daniel L. Mummey. "Mycorrhizas and soil structure." *New Phytologist* 171, no. 1 (2006): 41–53.
17. Rousk, Johannes, and Erland Bååth. "Growth of saprotrophic fungi and bacteria in soil." *FEMS Microbiology Ecology* 78, no. 1 (2011): 17–30.
18. Wu, Di, Mengmeng Zhang, Mu Peng, et al. "Variations in soil functional fungal community structure associated with pure and mixed plantations in typical temperate forests of China." *Frontiers in Microbiology* 10 (2019): 1636.
19. Jacoby, Richard, Manuela Peukert, Antonella Succurro, et al. "The role of soil microorganisms in plant mineral nutrition: Current knowledge and future directions." *Frontiers in Plant Science* 8 (2017): 1617.
20. Naveed, Muhammad, Lasantha Herath, Per Moldrup, et al. "Spatial variability of microbial richness and diversity and relationships with soil organic carbon, texture and structure across an agricultural field." *Applied Soil Ecology* 103 (2016): 44–55.
21. Luo, Xia, Kezhong Liu, Yuyu Shen, et al. "Fungal community composition and diversity vary with soil horizons in a subtropical forest." *Frontiers in Microbiology* 12 (2021).
22. Jiao, Shuo, Weimin Chen, Jieli Wang, et al. "Soil microbiomes with distinct assemblies through vertical soil profiles drive the cycling of multiple nutrients in reforested ecosystems." *Microbiome* 6, no. 1 (2018): 1–13.
23. Yan, Hongmei, Fan Yang, Jiamin Gao, et al. "Subsoil microbial community responses to air exposure and legume growth depend on soil properties across different depths." *Scientific Reports* 9, no. 1 (2019): 1–11.
24. Liu, Nana, Huifeng Hu, Wenhong Ma, et al. "Contrasting biogeographic patterns of bacterial and archaeal diversity in the top-and subsoils of temperate grasslands." *MSystems* 4, no. 5 (2019): e00566-19.

25. Erktan, Amandine, Dani Or, and Stefan Scheu. "The physical structure of soil: determinant and consequence of trophic interactions." *Soil Biology and Biochemistry* 148 (2020): 107876.
26. Kim, Hye Min, Min Jin Lee, Ji Young Jung, et al. "Vertical distribution of bacterial community is associated with the degree of soil organic matter decomposition in the active layer of moist acidic tundra." *Journal of Microbiology* 54, no. 11 (2016): 713–723.
27. Keshavarzi, Ali, Henry Oppong Tuffour, Ali Bagherzadeh, et al. "Spatial and fractal characterization of soil properties across soil depth in an agricultural field, Northeast Iran." *Eurasian Journal of Soil Science* 7, no. 2 (2018): 93–102.
28. Feng, Huili, Jiahuan Guo, Weifeng Wang, et al. "Soil depth determines the composition and diversity of bacterial and archaeal communities in a poplar plantation." Forests 10, no. 7 (2019): 550.
29. Xiao, Ye, Zhigang Huang, and Xianguo Lu. "Changes of soil labile organic carbon fractions and their relation to soil microbial characteristics in four typical wetlands of Sanjiang Plain, Northeast China." *Ecological Engineering* 82 (2015): 381–389.
30. Naylor, Dan, Ryan McClure, and Janet Jansson. "Trends in microbial community composition and function by soil depth." *Microorganisms* 10, no. 3 (2022): 540.
31. Fierer, Noah, Joshua P. Schimel, and Patricia A. Holden. "Variations in microbial community composition through two soil depth profiles." *Soil Biology and Biochemistry* 35, no. 1 (2003): 167–176.
32. Hoorman, J.J. *Understanding Soil Microbes and Nutrient Recycling: SAG-16, Agriculture and Natural Resources.* Ohio State University, Ohio, 2010.
33. Fierer, Noah, Mark A. Bradford, and Robert B. Jackson. "Toward an ecological classification of soil bacteria." *Ecology* 88, no. 6 (2007): 1354–1364.
34. Bhattarai, A., B. Bhattarai, and S. Pandey. "Variation of soil microbial population in different soil horizons." *J Microbiol Exp* 2, no. 2 (2015): 00044.
35. Hamarashid, Najmadeen H., Mohammad A. Othman, and Mohamed-Amin H. Hussain. "Effects of soil texture on chemical compositions, microbial populations and carbon mineralization in soil." *Egyptian Journal of Experimental Biology (Botany)* 6, no. 1 (2010): 59–64.
36. Wick, Abbey F., Snehalata V. Huzurbazar, and Peter D. Stahl. "Use of Bayesian methods to model soil aggregation in undisturbed semiarid grasslands." *Soil Science Society of America Journal* 73, no. 5 (2009): 1707–1714.
37. Knops, Johannes M. H., and David Tilman. "Dynamics of soil nitrogen and carbon accumulation for 61 years after agricultural abandonment." *Ecology* 81, no. 1 (2000): 88–98.
38. Chau, Jessica Furrer, Amvrossios C. Bagtzoglou, and Michael R. Willig. "The effect of soil texture on richness and diversity of bacterial communities." *Environmental Forensics* 12, no. 4 (2011): 333–341.
39. Maeght, Jean-Luc, Boris Rewald, and Alain Pierret. "How to study deep roots: And why it matters." *Frontiers in Plant Science* 4 (2013): 299.
40. Bach, Elizabeth M., Sara G. Baer, Clinton K. Meyer, and Johan Six. "Soil texture affects soil microbial and structural recovery during grassland restoration." *Soil Biology and Biochemistry* 42, no. 12 (2010): 2182–2191.
41. Whiteside, Matthew D., Gijsbert D. A. Werner, Victor E. A. Caldas, et al. "Mycorrhizal fungi respond to resource inequality by moving phosphorus from rich to poor patches across networks." *Current Biology* 29, no. 12 (2019): 2043–2050.
42. Geisseler, Daniel, and Kate M. Scow. "Long-term effects of mineral fertilizers on soil microorganisms: A review." *Soil Biology and Biochemistry* 75 (2014): 54–63.

43. Bending, Gary D., Mary K. Turner, and Julie E. Jones. "Interactions between crop residue and soil organic matter quality and the functional diversity of soil microbial communities." *Soil Biology and Biochemistry* 34, no. 8 (2002): 1073–1082.
44. Hao, X. H., S. L. Liu, J. S. Wu, et al. "Effect of long-term application of inorganic fertilizer and organic amendments on soil organic matter and microbial biomass in three subtropical paddy soils." *Nutrient Cycling in Agroecosystems* 81, no. 1 (2008): 17–24.
45. Cookson, William R., Daniel A. Abaye, Petra Marschner, et al. "The contribution of soil organic matter fractions to carbon and nitrogen mineralization and microbial community size and structure." *Soil Biology and Biochemistry* 37, no. 9 (2005): 1726–1737.
46. Bhandari, Krishna B., Charles P. West, and Veronica Acosta-Martinez. "Assessing the role of interseeding alfalfa into grass on improving pasture soil health in semi-arid Texas High Plains." *Applied Soil Ecology* 147 (2020): 103399.
47. Brady, A. C., and R. R. Weil *The Nature and Properties of Soils*. 13th ed. Prentice Hall, New Jersey, 2002.
48. Zitnick-Anderson, Kimberly K., Jack E. Norland, et al. "Probability models based on soil properties for predicting presence-absence of *Pythium* in soybean roots." *Microbial Ecology* 74, no. 3 (2017): 550–560.
49. Andersson, Stefan, S. Ingvar Nilsson, and Peter Saetre. "Leaching of dissolved organic carbon (DOC) and dissolved organic nitrogen (DON) in mor humus as affected by temperature and pH." *Soil Biology and Biochemistry* 32, no. 1 (2000): 1–10.
50. Pietri, J. C. Aciego, and P. C. Brookes. "Relationships between soil pH and microbial properties in a UK arable soil." *Soil Biology and Biochemistry* 40, no. 7 (2008): 1856–1861.
51. Marstorp, Håkan, Xin Guan, and Ping Gong. "Relationship between dsDNA, chloroform labile C and ergosterol in soils of different organic matter contents and pH." *Soil Biology and Biochemistry* 32, no. 6 (2000): 879–882.
52. Fierer, Noah, and Robert B. Jackson. "The diversity and biogeography of soil bacterial communities." *Proceedings of the National Academy of Sciences* 103, no. 3 (2006): 626–631.
53. Jiang, Mingyi, and Jianhua Zhang. "Water stress-induced abscisic acid accumulation triggers the increased generation of reactive oxygen species and up-regulates the activities of antioxidant enzymes in maize leaves." *Journal of Experimental Botany* 53, no. 379 (2002): 2401–2410.
54. Giacometti, Caterina, Michael Scott Demyan, Luciano Cavani, et al. "Chemical and microbiological soil quality indicators and their potential to differentiate fertilization regimes in temperate agroecosystems." *Applied Soil Ecology* 64 (2013): 32–48.
55. Unger, Irene M., Ann C. Kennedy, and Rose-Marie Muzika. "Flooding effects on soil microbial communities." *Applied Soil Ecology* 42, no. 1 (2009): 1–8.
56. Herron, Patrick M., John M. Stark, Carson Holt, et al. "Microbial growth efficiencies across a soil moisture gradient assessed using 13C-acetic acid vapor and 15N-ammonia gas." *Soil Biology and Biochemistry* 41, no. 6 (2009): 1262–1269.
57. Fierer, Noah, and Joshua P. Schimel. "A proposed mechanism for the pulse in carbon dioxide production commonly observed following the rapid rewetting of a dry soil." *Soil Science Society of America Journal* 67, no. 3 (2003): 798–805.
58. Yan, Nan, Petra Marschner, Wenhong Cao, et al. "Influence of salinity and water content on soil microorganisms." *International Soil and Water Conservation Research* 3, no. 4 (2015): 316–323.

59. Jin, Virginia L., Richard L. Haney, Philip A. Fay, and H. Wayne Polley. "Soil type and moisture regime control microbial C and N mineralization in grassland soils more than atmospheric CO_2-induced changes in litter quality." *Soil Biology and Biochemistry* 58 (2013): 172–180.
60. Gruber, Nicolas, and James N. Galloway. "An Earth-system perspective of the global nitrogen cycle." *Nature* 451, no. 7176 (2008): 293–296.
61. Erisman, Jan Willem, James N. Galloway, Sybil Seitzinger, et al. "Consequences of human modification of the global nitrogen cycle." *Philosophical Transactions of the Royal Society B: Biological Sciences* 368, no. 1621 (2013): 20130116.
62. Berthrong, Sean T., Chris M. Yeager, Laverne Gallegos-Graves, Blaire Steven, Stephanie A. Eichorst, Robert B. Jackson, and Cheryl R. Kuske. "Nitrogen fertilization has a stronger effect on soil nitrogen-fixing bacterial communities than elevated atmospheric CO_2." *Applied and Environmental Microbiology* 80, no. 10 (2014): 3103–3112.
63. Liu, Man, Guilin Han, and Xiaoqiang Li. "Using stable nitrogen isotope to indicate soil nitrogen dynamics under agricultural soil erosion in the Mun River basin, Northeast Thailand." *Ecological Indicators* 128 (2021): 107814.
64. Tian, Dashuan, and Shuli Niu. "A global analysis of soil acidification caused by nitrogen addition." *Environmental Research Letters* 10, no. 2 (2015): 024019.
65. Bowman, William D., Leslie Bahn, and Mary Damm. "Alpine landscape variation in foliar nitrogen and phosphorus concentrations and the relation to soil nitrogen and phosphorus availability." *Arctic, Antarctic, and Alpine Research* 35, no. 2 (2003): 144–149.
66. Bracken, Matthew E. S., Helmut Hillebrand, Elizabeth T. Borer, et al. "Signatures of nutrient limitation and co-limitation: Responses of autotroph internal nutrient concentrations to nitrogen and phosphorus additions." *Oikos* 124, no. 2 (2015): 113–121.
67. Houben, David, Etienne Michel, Cécile Nobile, et al. "Response of phosphorus dynamics to sewage sludge application in an agroecosystem in northern France." *Applied Soil Ecology* 137 (2019): 178–186.
68. Robbins, Chanz, Thorsten Thiergart, Stéphane Hacquard, et al. "Root-associated bacterial and fungal community profiles of Arabidopsis thaliana are robust across contrasting soil P levels." *Phytobiomes* 2, no. 1 (2018): 24–34.
69. Ducousso-Détrez, Amandine, Joël Fontaine, Anissa Lounès-Hadj Sahraoui, and Mohamed Hijri. "Diversity of phosphate chemical forms in soils and their contributions on soil microbial community structure changes." *Microorganisms* 10, no. 3 (2022): 609.

4 Soil microbiome of cold deserts and its biotechnologically useful adaptations for life on Earth and beyond, with a detailed case study on the Atacama Desert microbiome

*Harshal Wasnik, Rishi Gandhi, Lucinda Elizabeth Doyle and Ishaan Gupta**

CONTENTS

* Corresponding author e-mail: ishaan@iitd.ac.in

DOI: 10.1201/9781003354031-4

4.1 INTRODUCTION

In 1902, Schmidt-Nielsen grew microorganisms at 0 °C, which he termed 'psychrophiles'. *Vibrio* (*Moritella* gen. nov.) *marinus* (*marina* comb. nov.) MP-1 and *Vibrio* (*Colwellia* gen. nov.) were the first true psychrophiles to be taxonomically described in the literature.[1] Current genomic-based studies suggest that the most transcriptionally active phyla in the cold environment are Proteobacteria, Firmicutes, Acidobacteria and Actinobacteria, as well as Euryarchaeota and ascomycetous fungi.[2]

The first set of adaptations in cryophiles is to protect against damage to biomolecules in colds, such as dedicated DNA damage repair mechanisms and protein misfolding by antifreeze proteins and isozymes. Isozymes are used to prevent cold protein denaturation and to maintain stability at a lower temperature. Prokaryotic species are also known to synthesise cryoprotectants (CRPs), which are organic compounds that accumulate in the body fluids.[3] Trehalose, glycerol and sorbitol are popular CRPs that aid in the anti-freezing process when exposed to sub-zero temperatures. The second collection of adaptations allows efficient biochemistry and cell biology to occur effectively at lower temperatures through changes in the composition and structure of various biomolecules found in cryophiles. Cryophiles have high glycine residues, low proline and low arginine content in their proteins, leading to a reduction in the number of ion-pairs, hydrogen bonding, and hydrophobic activity (Figure 4.1).[4] Adaptation also involves a change in the membrane fatty acid content, which correlates with unsaturation, an increase/decrease in chain length branching.[5]

4.2 CRYOPHILES IN COLD DESERTS OF THE WORLD

Deserts or drylands cover more than one-third of the Earth's landmass. The desert soil is characterised by low bioavailability of water due to the near-complete absence of precipitation, which is further exacerbated by high evaporation in the case of hot deserts or sublimation in the case of cold deserts. In other challenges, these areas also experience high variation in diurnal and seasonal changes in surface temperature, imposing heat and cold shocks, and receive markedly higher penetration of solar radiations. Therefore, besides the previously described adaptations, the cryophiles constituting the microbiome of cold deserts have been demonstrated or are expected to have cellular adaptations for photoprotection by producing UV-absorbing pigments such as melanins or scytonemin,[6] adaptations to UV damage to DNA by scavenging of reactive oxygen species through

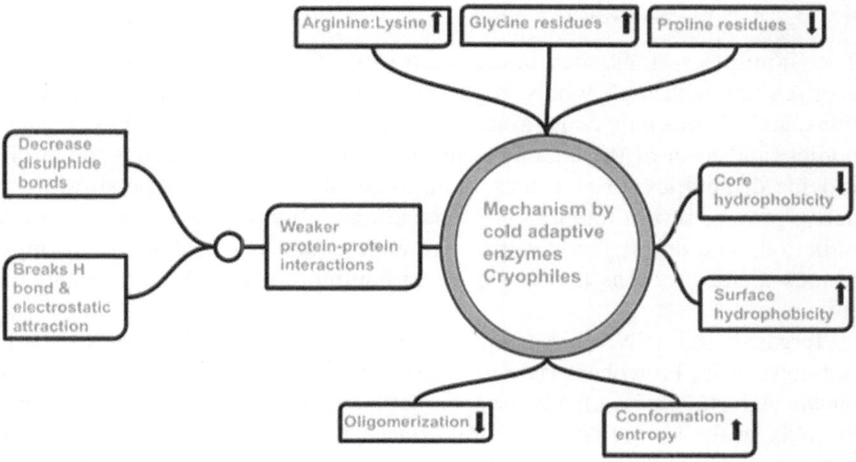

FIGURE 4.1 Various mechanisms of cold-adaptive enzymes in cryophiles.

manganese complexes, and desiccation tolerance by producing molecules such as trehalose and modifications to the cell membrane or cell walls.[7] The following subsections review the microbiology of five such cold desert habitats around Asia, Africa, and South America.

4.2.1 The Gobi Desert

The Gobi Desert is a cold desert, ranked the fifth largest in the world and most extensively studied in the Asian subcontinent. The desert has subsets of arid regions and is mainly subjected to high UV radiation, low organic matter and temperature fluctuations. Targeted 16S rRNA amplicon sequencing indicates more than 15 different phyla present in the desert, with Firmicutes, Proteobacteria, Bacteroidetes and Actinobacteria the most abundant phyla, accounting for 90% of the microbial diversity present in the desert.[8] The characterisation of the fungal population reports an abundance of the Ascomycota phylum from the Dothideomycetes and Sordariomycetes families, while Basidiomycetes were the second most abundant in the total readings.

4.2.2 The Taklamakan Desert

The Taklamakan Desert is located in Xinjiang, north-western China, and is limited to the west by mountains and to the east by the Gobi Desert. Actinobacteria, Proteobacteria, Bacteroidetes and Firmicutes have been found in abundance in desert studies.[6] Radiation- and desiccation-tolerant bacteria have been identified over the years, which can also accumulate high Pb content. The bacteria found here are especially resistant to desiccation.

4.2.3 The Namib Desert

The Namib Desert, situated in the south-west of Africa, is a subset of hyper-arid regions with high winds, high UV intensity, wide temperature fluctuations, and oligotrophic constraints in the soil ecosystems. Namib has no water bodies, and most of the channels run underground. Poly-extreme conditions provide difficulties for soil microbiota to adapt to elevated xeric stress (oxidative stress), severe temperature fluctuations, oligotrophy and extreme UV radiation. The desert has diverse communities and is often limited to mysterious niches, such as hypoliths, where communities are located under the stones.

Targeted 16S rRNA amplicon sequencing reported similar results, with Actinobacteria, Proteobacteria and Cyanobacteria the most prevalent bacterial phyla.[9] A further class of Ascomycota, dothideomycetes, dominated the fungal diversity of the Namib desert.[10] Viral sequence analysis studies have also been conducted in this desert, where the most prevalent species were Caudovirales, Phycodnaviradae and Mimiviridae.[11]

4.2.4 The Drass Desert

Drass is known to be India's coldest. and the world's second coldest, human-populated location after Siberia. The desert has subsets of soil habitats of different conditions that mainly involve considerable variations in temperature and low organic content. Drass soil is slightly alkaline and sandy, and has electrical conductivity (EC). Various metagenomics studies, like culture-dependent approaches, report various Proteobacteria, Actinobacteria and Firmicutes.[12] Various pigmented bacteria were also reported in the study. The pigments were reported to be carotenoids produced in response to various stresses such as low temperature in cold deserts. Targeted 16sRNA amplicon sequencing reported the abundance of Acidobacteria, Proteobacteria and Actinobacteria. Other phyla, such as Bacteroidetes, Cyanobacteria, Chloroflexi, Firmicutes, Gemmatimonadetes, Planctomycetes, Nitrospira and Verrucomicrobia, were also reported at pretty low density.[10]

4.2.5 The Atacama Desert

Located in South America, the Atacama Desert is known to be one of the world's driest places. In most of its parts, the desert has extremely arid areas and almost no rain. The Atacama Desert displays various extremes, such as extreme day–night variations, the most arid conditions on Earth, oxidising environments and the highest UV radiation measured on Earth.[13] Such conditions make it the ideal candidate for work on terraforming planets like Mars. The Atacama Desert has a unique environment and can respond to the unresolved 'life-limit' question. The desert has unique bacterial diversity; Actinobacteria, together with Proteobacteria, Acidobacteria and Chloroflexi, are the predominant species in these areas.[14] The Proteobacteria are the most abundant bacteria in the Atacama, accounting for 18% of the total microbiome, which is almost twice as much as the Actinobacteria present.[13] Other bacterial species, including species of the genera *Bacillus, Arthrobacter, Brevibacillus, Cellulomonas and Hymenobacter,* have also been recovered in the Atacama. The Atacama Desert recorded its first novel cyanobacteria, *Calothrix aesertica, Schizothrix adunca* and *Schizothrix atacamensis,* around 50 years ago.

4.3 ELECTROACTIVE CRYOPHILES

4.3.1 Introduction to Extracellular Electron Transfer

Electroactive microorganisms are capable of producing microbial electricity in a remarkable form of metabolism known as extracellular electron transfer. Microorganisms achieve this by exchanging electrons with solid conductive surfaces (e.g., metals in nature and electrodes in a laboratory).[15] Such surfaces typically act as the organism's terminal electron acceptor,[16] in stark contrast to the soluble acceptors once thought essential for life. In pursuit of solutions to the modern energy crisis, electroactive microorganisms have been deployed in a variety of bioelectrochemical devices to harness their sustainable, environmentally friendly electrical current. Even before extracellular electron transfer was discovered, NASA conducted a number of studies in the 1960s with the goal of using microorganisms to bioelectrochemically generate power from human waste in space vehicles.[17] More modern efforts were established with the development of the microbial fuel cell, which focuses on the extracellular electron transfer process. With such a device, the goal is typically to treat wastewater using electroactive microorganisms while simultaneously producing power.[18]

To date, three mechanisms for transporting electrons across the cell envelope have been discovered: via redox-active proteins (outer-membrane cytochromes), via conductive microbial appendages (nanowires and e-pili) and via soluble electron shuttles such as riboflavin (either endogenously produced or exogenously added).[19] The study of such fundamentals of electroactive microorganisms has evolved into the field of electromicrobiology. While the majority of electroactive microorganisms discovered have been bacteria (including certain cyanobacteria), extracellular electron transfer has also been observed in archaea and yeasts.[20, 21] In addition to mechanism and diversity, the ecology of these microorganisms is also an area of investigation in electromicrobiology. Conventionally, extracellular electron transfer was viewed to be a curiosity of environmental microbes inhabiting anaerobic soils and sediments and playing a role in biogeochemical cycling.[22] Recently, this perspective has been demonstrated to be too narrow, as medically relevant species were found capable of producing electricity, including *Enterococcus faecalis*[23, 24] and *Listeria monocytogenes*,[25] posing new questions about how and why microorganisms become electroactive. With these and other related discoveries, it has become clear that in nature, electroactivity exists along a spectrum, with microbes ranging from weak to strong electricigens depending on their extracellular electron transfer abilities.[26]

In light of their application in bioelectrochemical systems and their intriguing ecology, the enrichment and isolation of electroactive species from various environments has been underway in recent years, leading to a number of interesting discoveries, including those relevant to cold deserts.

4.3.2 Electroactive Microbes in the Gobi Desert

A plethora of electroactive species associated with desert varnish (also known as rock varnish) in the Gobi Desert has been discovered.[27] Desert varnish, a thin layer that forms naturally on exposed rock surfaces in arid regions, predominantly consists of clay and is rich in iron and manganese oxides.[28] Varnish-like structures have been observed on Mars,[29] drawing astrobiological attention to these geological structures on Earth. Due to electroactive microorganisms' proclivity to reduce metal oxides such as iron and manganese, rock varnishes are an excellent environment to probe for extracellular electron transfer activity in the natural environment. Table 4.1 details the electroactive genera associated with desert varnish in the Gobi Desert, as determined via culture-independent 16S rRNA gene profiling.[27] It was found that the varnish was heavily enriched with electroactive members as compared with the surrounding soil. The majority are Gram-negative, as is typical for electroactive microbes, but there are a few Gram-positive genera in the varnish community. A number of the Gobi Desert varnish electroactive community members are known weak electricigens, e.g., *Escherichia*, *Bacillus* and *Pseudomonas*.[26]

A comprehensive search for electroactive microbes in cold deserts has yet to be conducted outside the Gobi Desert. The Gobi Desert varnish study provides an intriguing glimpse into the dynamic interplay that may exist between electroactive microorganisms and their natural environment, highlighting the potential benefits of continuing such work.

4.4 ATACAMA DESERT: A CASE STUDY FOR BIOTECHNOLOGY ON EARTH AND BEYOND

The Atacama Desert introduces us to a group of cryophiles with the potential to be used for bioremediation and producing bioactive natural products. Researchers have documented the various bacterial species that are capable of bioremediation. There has been taxonomic study of distinct Streptomyces clades that has contributed to the identification of bioactive compounds that display promising antibacterial and anticancer properties. The bacteria found in Atacama can withstand massive metal toxicity and have mineral binding affinities, which are attributed to leaching properties, therefore acting as an excellent bioleaching agent.

TABLE 4.1
Electroactive genera discovered in the Gobi Desert.[27] Details of each entry's extracellular electron transfer (ascertained from separate studies) are included

Electroactive genus	Nature of extracellular electron transfer (in bioelectrochemical device)	Cell envelope
Acinetobacter	Can use electrode as electron donor,[35] possibly via endogenously produced redox shuttle[36]	Gram –ve
Brevundimonas	May use electrode as electron donor[35] via unknown mechanisms. May also use electrodes as electron acceptor[37]	Gram –ve
Cronobacter	Can use electrode as electron acceptor via unknown mechanism[38]	Gram –ve
Escherichia	Can use electrodes as electron donor via redox shuttles; primarily exogenously added.[39] May also use electrodes as electron donors[35]	Gram –ve
Ochrobactrum	Can use electrode as electron acceptor,[40] possibly via endogenously produced soluble redox shuttle[41]	Gram –ve
Pseudomonas	Can use electrode as electron acceptor via endogenously produced redox shuttles (phenazines).[42] May also use electrode as electron donor[35]	Gram –ve
Stenotrophomonas	Can use electrode as electron acceptor via redox shuttles (exogenously added)[43]	Gram –ve
Bacillus	Can use electrode as electron acceptor via endogenously produced redox shuttle (flavins).[44] May also use electrode as electron donor[35]	Gram +ve
Dietzia	Can use electrode as electron acceptor via endogenously produced redox shuttle (the carotenoid canthaxanthin)[45]	Gram +ve
Staphylococcus	Can use electrode as electron acceptor, possibly via a direct mechanism that does not depend on soluble redox shuttles.[46] May also use electrodes as electron donors[35]	Gram +ve

In Table 4.2, we summarise the key strains found in the Atacama Desert along with their potential use in various biotechnology applications.

4.4.1 Exobiology for life beyond Earth: Do we have Mars on Earth?

The Atacama Desert has received significant attention from the scientific community since the mid-1950s. NASA was a pioneer in studying the Atacama Desert. Low levels of organic matter, the presence of oxidising agents (which can oxidise l- and d-amino acids), and oligotrophic nutrient level in soils

TABLE 4.2
Biotechnology potential of microbes isolated from Atacama Desert

Microbe(s)	Biotechnology Application
Pseudomonas arsenicoxydans strain VC-1	Bioremediation of arsenic[47]
Streptomyces sp. Lt 005, *Streptomyces* C1 and *Streptomyces* sp. CBS 198.65	Potential new drugs for the avoidance of neurodegenerative disorders such as Parkinson's and Alzheimer's disease by molecules that scavenge reactive oxygen species[48]
Streptomyces sp. Strain DB634	Synthesis of secondary metabolites, called abenquines (A–D), that show inhibitory activity against *Bacillus subtilis* and anti-fibroblast proliferation[49]
Streptomyces leeuwenhoekii strain C34 co-cultivation with *Aspergillus fumigatus* MR2012	Produces luteoride D and induces pseurotin G, which has anticancer bioactivity[50]
Streptomyces leeuwenhoekii strain C58 co-cultivation with *Aspergillus fumigatus* MR2012	Produces metabolites pentalenic acid and chaxapeptin (enhanced titre, almost double), which have anticancer bioactivity[49]
Fungus UFMGCB 8046 sequence similarity with *Aspergillus persii*	Cytotoxic activity against tumoral cell line MCF-7[50]
Fungal strain UFMGCB 8023, identified as *F. oxysporum*	Cytotoxic activity against tumoral cell line MCF-7[50]
Fungus *Hypoxylon* sp. UFMGCB 8020	Antiviral activity against dengue virus[50]
Fungus FP. *chrysogenum* UFMGCB 8074	α-linoleic acid and ergosterol endoperoxide (fungal and bacterial targets)[50]
Streptomyces sp. strain C34	Isolation and identification of three new polyketide compounds, named chaxalactins A–C, that have antibacterial and immunosuppressive properties[51]
Streptomyces sp. C38	Three new macrolactone antibiotics (Atacamycins A–C) that showed inhibitory activity against enzyme phosphodiesterase (PDE-4B2) and are therefore potential therapeutics for coronary heart disease, dementia, depression and schizophrenia[52]
Leptospirillum ferrooxidans and *Ferroplasma acidiphilum*	Bioleaching of ferrous iron oxidation and nitrogen[53]
Sulfobacillus sp., *Alicyclobacillus* sp. and *Acidiphilium* sp.	Acidophilic iron(II)- and sulphur-oxidising and iron(III)-reducing agents[54]

render the Atacama an ideal example to be named as the Martian subsurface equivalent.

In the late 1960s, Caltech Jet Propulsion Laboratory started work in the hyper-arid regions of the Atacama Desert. Initial studies on salt flats and sand dunes around Antofagasta indicate the presence of Actinomycetes and aerophilic bacteria.[30] Further studies were carried out in 1975 to develop the NASA Viking Mission Life Detection System. Because of the growing interest in it as a kind of

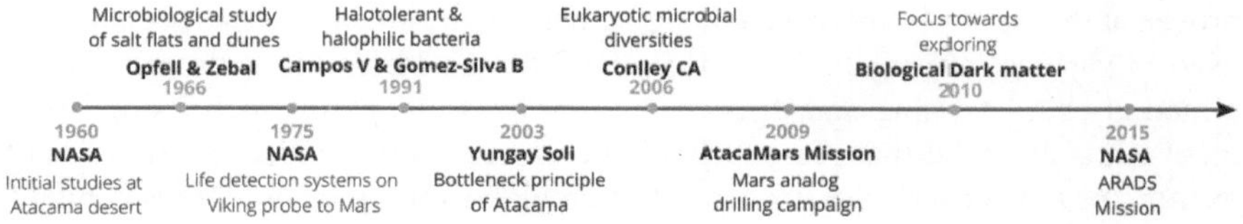

FIGURE 4.2 Timeline of studies conducted in the Atacama Desert.

extraterrestrial land, a number of studies[31] in different regions of Atacama have been carried out with keen interest.

Most microbes in Atacama that survive without water enter the state of anhydrobiosis, where cells stop metabolising but remain viable. Therefore, the detection of such microbes becomes difficult and challenging. Emerging metagenomics techniques such as Oxford Nanopore provide a solution to this problem. Search for Extra-Terrestrial Genomes (SETG) may be a possible way to determine life's existence. Long-read nanopore sequencing can detect SETG containing non-standard bases and preserved regions. There are reports where methods for extracting nucleic acid from low biomass and Martian-like surfaces have been validated.[32, 33] Recently, NASA started the Atacama Rover Astrobiology Drilling Studies (ARADS) project, which aims to mimic the feasibility of roving and drilling missions to Mars in search of life (Figure 4.2).[34]

Further input into the Atacama Desert continues to be relevant to the study of extraterrestrial life signatures, and efforts such as the establishment of an Astrobiology Field Laboratory for Geobiology Studies have aided in the cause.

4.5 CONCLUSION AND FUTURE OF COLD SOIL MICROBIOME RESEARCH: METAGENOMICS PERSPECTIVE

Here, we presented an updated comprehensive review of soil microbiomes of cold deserts around the world. We detailed the Atacama Desert microbiome, which is suitable for upcoming researchers interested in biotechnologically advanced extremozymes. We hope that this chapter will motivate further research into the Atacama as a potential hub of extremozymes and a place for future astrobiological research, such as space traveling, space farming and human hibernation.

As described in earlier sections, most studies selectively amplify highly conserved markers such as 16SrRNA from the soil samples followed by sequencing to identify the diversity of microbes. This helps to identify genus-level information about the microbial composition of the soils. However, with the decreasing cost of sequencing, researchers are quickly moving towards sequencing the entire contents of DNA extracted from the soil samples and developing computational tools that help them to identify even the species-level information from these samples.

Next-generation sequencing has generated so much data from extreme environments. The hand-held sequencing device called the Oxford Nanopore has been proven at the point of care in a challenging field environment and has been readily used in various investigations, such as the Extreme Microbiome Project (EMP) aimed at characterising and discovering novel extremophile pipelines and protocols. It is assumed that these technologies will help to predict an explosion of exciting new discoveries in the field of microbial diversity from various habitats, including cold deserts around the world.

ACKNOWLEDGEMENT

We thank Indian Institute of Technology Delhi and Indian Institute of Science Education and Research Bhopal for their support.

REFERENCES

1. Morita, Richard Y. "Psychrophilic bacteria." *Bacteriological Reviews* 39, no. 2 (1975): 144–167.
2. Coolen, Marco J. L., and William D. Orsi. "The transcriptional response of microbial communities in thawing Alaskan permafrost soils." *Frontiers in Microbiology* 6, no.197 (2015): 1–14.
3. Storey, Kenneth B. "Metabolism and bound water in overwintering insects." *Cryobiology* 20, no. 3 (1983): 365–379.
4. Perl, D., U. Mueller, U. Heinemann, and F. X. Schmid. 2000. "Two exposed amino acid residues confer thermostability on a cold shock protein." Nature Structural Biology 7 (5): 380–83.
5. Králová, Stanislava. "Role of fatty acids in cold adaptation of Antarctic psychrophilic *Flavobacterium* spp." *Systematic and Applied Microbiology* 40, no. 6 (2017): 329–333.
6. Gao, Qunjie, and Ferran Garcia-Pichel. 2011. "Microbial ultraviolet sunscreens." *Nature Reviews. Microbiology* 9 (11):791–802.
7. Crowe, John H., Folkert A. Hoekstra, and Lois M. Crowe. "Anhydrobiosis." *Annual Review of Physiology* 54 (1992): 579–599.
8. An, Shu, Cécile Couteau, Fan Luo, Julie Neveu, and Michael S. DuBow. 2013. "Bacterial diversity of surface sand samples from the Gobi and Taklamaken Deserts." Microbial Ecology 66 (4): 850–60.
9. Valverde, Angel, Thulani P. Makhalanyane, Mary Seely, and Don A. Cowan. 2015. "Cyanobacteria drive community composition and functionality in rock-soil interface communities." Molecular Ecology 24 (4): 812–21.
10. Walt, Andries J. van der, Riegardt M. Johnson, Don A. Cowan, et al. "Unique microbial phylotypes in namib desert dune and gravel plain fairy circle soils."*Applied and Environmental Microbiology* 82, no. 15 (2016): 4592–4601.
11. Adriaenssens, Evelien M., Lonnie Van Zyl, et al. "Metagenomic analysis of the viral community in namib desert hypoliths." *Environmental Microbiology* 17, no. 2 (2015): 480–95.
12. Gupta, Puja, Naseer Sangwan, Rup Lal, and Jyoti Vakhlu. "Bacterial diversity of drass, cold desert in Western Himalaya, and its comparison with Antarctic and Arctic." Archives of Microbiology 197, no. 6 (2015): 851–60.
13. Cordero, Raul R., Gunther Seckmeyer, et al. "The world's highest levels of surface UV." *Photochemical & Photobiological Sciences* 13, no. 1 (2013): 70–81.
14. Neilson, Julia W., Jay Quade, Marianyoly Ortiz, et al. "Life at the hyperarid margin: Novel bacterial diversity in arid soils of the Atacama Desert, Chile." *Extremophiles* 16, no. 3 (2012): 553–566.

15. Contador, Carolina A., Luis Veas-Castillo, Emilio Tapia, et al. "Atacama database: A platform of the microbiome of the Atacama desert." *Antonie Van Leeuwenhoek* 113, no. 2 (2020): 185–195.
16. Myers, C. R., and K. H. Nealson. "Bacterial manganese reduction and growth with manganese oxide as the sole electron acceptor." Science 240, no.4857 (1988): 1319–21.
17. Canfield, J. H., B. H. Goldner, and R. Lutwack. *Research on Applied Bioelectrochemistry First Quarterly Progress Report, 14 Mar.-30 Jun. 1963. No. NASA-CR-71834.* 1963.
18. Logan, Bruce E. *Microbial Fuel Cells.* John Wiley & Sons, 2008.
19. Shi, Liang, Hailiang Dong, Gemma Reguera, et al. "Extracellular electron transfer mechanisms between microorganisms and minerals." Nature Reviews. Microbiology 14, no.10 (2016): 651–62.
20. Hubenova, Yolina, and Mario Mitov. "Extracellular electron transfer in yeast-based biofuel cells: A review."Bioelectrochemistry 106, no. Pt A (2015): 177–85.
21. Koch, Christin, and Falk Harnisch. "Is there a specific ecological niche for electroactive microorganisms?" *Chem Electro Chem* 3, no. 9 (2016): 1282–1295.
22. Jiang, Yongguang, Meimei Shi, and Liang Shi. "Molecular underpinnings for microbial extracellular electron transfer during biogeochemical cycling of earth elements." Science China. Life Sciences 62, no.10 (2019): 1275–86.
23. Keogh, Damien, Ling Ning Lam, Lucinda E. Doyle et al. "Extracellular electron transfer powers *Enterococcus faecalis* biofilm metabolism." *MBio* 9, no. 2 (2018): e00626-17.
24. Pankratova, Galina, Dónal Leech, Lo Gorton, and Lars Hederstedt. "Extracellular electron transfer by the gram-positive bacterium *Enterococcus faecalis*." Biochemistry 57 no. 30 (2018): 4597–4603.
25. Light, Samuel H., Lin Su, Rafael Rivera-Lugo, et al. "A flavin-based extracellular electron transfer mechanism in diverse gram-positive bacteria." Nature 562, no. 7725 (2018): 140–44.
26. Doyle, Lucinda E., and Enrico Marsili. "Weak electricigens: A new avenue for bioelectrochemical research." Bioresource Technology 258 (2018): 354–64.
27. Ren, Guiping, Yingchun Yan, Yong Nie, et al "Natural extracellular electron transfer between semiconducting minerals and electroactive bacterial communities occurred on the rock varnish." *Frontiers in Microbiology* 10 (2019): 293.
28. Potter, R. M., and G. R. Rossman. "Desert Varnish: The importance of clay minerals." Science 196. no.4297 (1977): 1446–48.
29. Murchie, S., O. Barnouin-Jha, K. Barnouin-Jha, et al. "Old desert varnish-like coatings and young breccias at the Mars Pathfinder landing site." In Lunar and Planetary Science Conference, p. 1740. 2004.
30. Opfell, J. B., and G. P. Zebal. "Ecological patterns of micro-organisms in desert soils." Life Sciences and Space Research 5 (1967): 187–203.
31. Prado, B., A. Del Moral, E. Quesada et al. "Numerical taxonomy of moderately halophilic Gram-negative rods isolated from the Salar de Atacama, Chile." *Systematic and Applied Microbiology* 14, no. 3 (1991): 275–281.
32. Zabel, Robert H., Lyman W. Boomer, and Tom R. King. "A model of stress and burnout among teachers of behaviorally disordered students." *Behavioral Disorders* 9, no. 3 (1984): 215–221.
33. Parro, Victor, Graciela de Diego-Castilla, Mercedes Moreno-Paz, et al. "A microbial oasis in the hypersaline atacama subsurface discovered by a life detector chip: Implications for the search for life on Mars." *Astrobiology* 11, no.10 (2011): 969–96.

34. Glass, B., A. Davila, V. Parro, et al. "Atacama rover astrobiology drilling studies project: Second year." In *Earth and Space 2018: Engineering for Extreme Environments*, pp. 337–347. American Society of Civil Engineers, Reston, VA, 2018.
35. Cournet, Amandine, Marie-Line Délia, Alain Bergel, Christine Roques, and Mathieu Bergé. "Electrochemical reduction of oxygen catalyzed by a wide range of bacteria including Gram-positive." *Electrochemistry Communications* 12, no. 4 (2010): 505–508.
36. Freguia, Stefano, Seiya Tsujimura, and Kenji Kano. "Electron transfer pathways in microbial oxygen biocathodes." Electrochimica Acta 55, no. 3 (2010): 813–18.
37. Li, Hui, Bin Wang, Songping Deng, Jingcheng Dai, and Senlin Shao. "Oxygen-containing functional groups on bioelectrode surface enhance expression of c-type cytochromes in biofilm and boost extracellular electron transfer." *Bioresource Technology* 292 (2019): 121995.
38. Thulasinathan, Boobalan, Samsudeen Nainamohamed, James Obeth Ebenezer Samuel, et al. "Comparative study on *Cronobacter sakazakii* and *Pseudomonas otitidis* isolated from septic tank wastewater in microbial fuel cell for bioelectricity generation." Fuel 248 (2019): 47–55.
39. Park, D. H., and J. G. Zeikus. "Electricity generation in microbial fuel cells using neutral red as an electronophore." Applied and Environmental Microbiology 66, no. 4 (2000): 1292–97.
40. Zuo, Yi, Defeng Xing, John M. Regan, and Bruce E. Logan. "Isolation of the exoelectrogenic bacterium *Ochrobactrium anthropi* YZ-1 by using a U-tube microbial fuel cell." *Applied and Environmental Microbiology* 74, no.10 (2008): 3130–37.
41. Li, Xin, Guo-Zhen Zhong, Yan Qiao, et al. "A high performance xylose microbial fuel cell enabled by *Ochrobactrum* sp. 575 cells." *RSC Advances* 4, no. 75 (2014): 39839–43.
42. Wang, Yun, Suzanne E. Kern, and Dianne K. Newman. "Endogenous phenazine antibiotics promote anaerobic survival of *Pseudomonas seruginosa* via extracellular electron transfer." Journal of Bacteriology 192, no. 1 (2010): 365–69.
43. Galai, Said, Antonia Pérez de los Ríos, Francisco José Hernández-Fernández, et al. "Microbial fuel cell application for azoic dye decolorization with simultaneous bioenergy production using *Stenotrophomonas* sp." *Chemical Engineering & Technology* 38, no. 9 (2015): 1511–18.
44. Wu, Song, Yong Xiao, Lu Wang, et al. "Extracellular electron transfer mediated by flavins in gram-positive *Bacillus* sp. WS-XY1 and yeast *Pichia stipitis*." Electrochimica Acta 146 (2014): 564–67.
45. Sacco, Natalia J., M. Celina Bonetto, and Eduardo Cortón. "Isolation and characterization of a novel electrogenic bacterium, *Dietzia* sp. RNV-4." PloS One 12, no. 2 (2017): 2e0169955.
46. Bhuvaneswari, A., and Sheela Berchmans. "Metamorphosis of pathogen to electrigen at the electrode/electrolyte interface: Direct electron transfer of *Staphylococcus aureus* leading to superior electrocatalytic activity." *Electrochemistry Communications* 34 (2013): 25–28.
47. Campos, Victor L., Cristian Valenzuela, Pablo Yarza, et al. "*Pseudomonas arsenicoxydans* sp nov., an arsenite-oxidizing strain isolated from the Atacama desert." *Systematic and Applied Microbiology* 33, no. 4 (2010): 193–197.
48. Leiros, Marta, Eva Alonso, Jon A. Sanchez, et al. "Mitigation of ROS insults by *Streptomyces* secondary metabolites in primary cortical neurons." *ACS Chemical Neuroscience* 5, no. 1 (2014): 71–80.

49. Wakefield, Jennifer, Hossam M. Hassan, Marcel Jaspars, Rainer Ebel, and Mostafa E. Rateb. Dual induction of new microbial secondary metabolites by fungal bacterial co-cultivation. *Frontiers in Microbiology* 8 (2017): 1284.
50. Gonçalves, Vívian N., Charles L. Cantrell, David E. Wedge, et al. Fungi associated with rocks of the Atacama Desert: Taxonomy, distribution, diversity, ecology and bioprospection for bioactive compounds. *Environmental Microbiology* 18, no. 1 (2016): 232–245.
51. Rateb, Mostafa E., Wael E. Houssen, William TA Harrison, et al. "Diverse metabolic profiles of a *Streptomyces* strain isolated from a hyper-arid environment. *Journal of Natural Products* 74, no. 9 (2011): 1965–1971.
52. Nachtigall, Jonny, Andreas Kulik, Soleiman Helaly, et al. Atacamycins A–C, 22-membered antitumor macrolactones produced by *Streptomyces* sp. C38. *The Journal of Antibiotics* 64, no. 12 (2011): 775–780.
53. Merino, M. P., B. A. Andrews, and J. A. Asenjo. Stoichiometric model and flux balance analysis for a mixed culture of *Leptospirillum ferriphilum* and *Ferroplasma acidiphilum*. *Biotechnology Progress* 31, no. 2 (2015): 307–315.
54. Korehi, H., M. Blothe, M. A. Sitnikova, B. Dold, and A. Schippers. Metal mobilization by iron-and sulfur-oxidizing bacteria in multiple extreme mine tailings in the Atacama Desert, Chile. *Environmental Science & Technology* 47, no. 5 (2013): 2189–2196.

5 Phosphorus, potassium and iron solubilising bacteria in trans-Himalayan soil

*Neha Rani Bhagat, Krishna Kumar**
Somen Acharya and Vijay K. Bharti

CONTENTS

5.1 INTRODUCTION

The trans-Himalayan regions are remote cold desert stretches, characterised by the complex interplay of climate including prolonged sub-zero temperatures, low annual precipitation, low humidity (<30%), intense solar radiation, hypobaric-hypoxia, rugged terrain and limited natural resource availability.[1] These are scarcely populated regions, where the local population totally depend on agriculture and animal husbandry for their livelihood.[2] Since the population in these

* Corresponding author e-mail: drkrishna02@gmail.com

DOI: 10.1201/9781003354031-5

regions is continuously growing, more agricultural productivity is needed to meet food and nutritional demands, emphasising the importance of sustainable agriculture in these regions.[3]

Recently, it has been proclaimed that the microorganisms of soil play a major role in determining nutrient availability, soil health and crop productivity by regulating nutrient cycling and the fertility of the soil ecosystem.[4] These beneficial microbes are the free-living organisms colonising the rhizoplane, phyllosphere and rhizosphere, which positively affect the soil and plant productivity by directly providing the essential nutrients: phosphate, potassium and iron.[5] Hence, the best way studied for augmenting plant productivity and nutrient richness is through increasing the availability of essential mineral nutrients to plants with the help of plant growth promoting microbes (PGPMs) (phosphate, potassium and iron solubilising microorganisms).[5]

In this context, the microbial diversity of the trans-Himalayan regions is already known for its unique adaptive strategies, bioremediating properties and involvement in various biogeochemical cycles.[6] These properties make it a promising candidate for improving agro-environmental sustainability. In fact, the utilisation of indigenous PGPMs appears to be an advantageous tactic for the sustainable production of plants in such a stressful environment.[7–9] However, very limited reports are available on the diversity of these beneficial microbes and their interactions with soil and plants. Therefore, in the present chapter, an attempt was made to review various research findings for a better understanding of the studied trans-Himalayan microorganisms, which will help select potential PGPMs that can function optimally under specific environmental conditions for beneficial outcomes.

5.2 PLANT GROWTH PROMOTING MICROBES (PGPMS)

Many studies have revealed that soil health is highly influenced by the status of nutrition in soils and microbial activity, where soil microorganisms play a major part in the regulation of soil fertility, texture, mineral cycling and bioavailability of essential nutrients (iron, nitrogen, phosphorus and potassium).[4] In soil, these beneficial microbes are also designated as plant growth promoting microbes. PGPMs augment the growth of plants, usually through an increase in the acquisition of resources, nitrogen fixation rate and nutrient availability rate, as illustrated in Figure 5.1.

These microbes regulate plant growth by enhancing micro- and macronutrient intake. PGPMs decrease the soil pH through organic acid production or by stimulation of the proton pump (ATPase), leading to an improvement in the solubilisation of these nutrients. For example, plants cannot absorb the major proportion of phosphorus directly due to its insoluble nature, thereby causing limitation in their growth. Therefore, these PGPMs transform the insoluble form into a soluble form of phosphorus (HPO_4^{2-}/PO_4^{-}). Recently, Yadav et al.[10] showed that soil samples from Himalayan cold deserts are rich in microbial species of phyla 41% Firmicutes, 37% Proteobacteria, 19% Actinobacteria and 3% Bacteroidetes.[10]

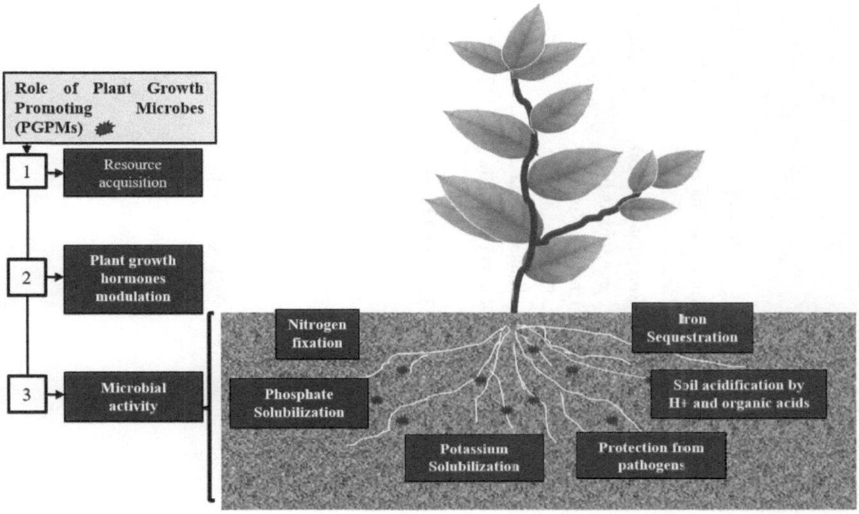

FIGURE 5.1 Schematic representation of PGPMs' role in plant growth promotion.

Moreover, PGPMs isolated and studied from trans-Himalayan regions are discussed in Table 5.1.

5.3 LABORATORY METHODOLOGY FOR SAMPLING AND ISOLATION OF PHOSPHATE, POTASSIUM AND IRON SOLUBILISING MICROORGANISMS

PGPMs mostly reside in the free-living soil, rhizosphere, rhizoplane and phyllosphere of the plant. So, soil samples are usually collected from the rhizoplane, rhizosphere and edaphosphere (non-rhizospheric zones).[25] Rhizospheric soil can be collected by uprooting the plant, rhizoplane soil by rinsing the immediate surface of roots in sterile normal saline solution, and leaves for the phyllosphere.[26] Then, all the collected samples should be kept in sampling bags and preserved at 4 °C. Further, the isolation of organisms can be done based on serial dilution plating or by the enrichment culturing method using various synthetic media (e.g., nutrient agar media, Luria Bertani media, etc., for the cultivation of bacteria and potato dextrose agar media, Sabouraud dextrose agar media, etc., for the cultivation of fungi).[14,27] After incubation, the grown colonies will be differentiated by their phenotypic colonial features and purified by repeated streaking.[25,28]

Thereafter, the potential isolates having phosphate, potassium and iron solubilising activities can be screened and selected by analysing the clear zones formed surrounding the colonies in Pikovskaya's (PVK) agar (a selective medium for phosphate), modified Aleksandrov medium containing potassium minerals[23,29]

TABLE 5.1
Nutrient solubilising microbes found in the trans-Himalayan region

Sr. No.	Microorganism	Nutrient	Region	References
1.	*Bacillus*, *Pseudomonas*, *Serratia*, *Enterobacter*, *Micrococcus*, *Flavobacterium*, *Penicillium*, *Aspergillus fumigatus*, *Aspergillus niger*	Phosphate, potassium	Himachal Pradesh, India	11,12
2.	*Pseudomonas* spp: *Pseudomonas trivialis*, *P. poae*, *P. fluorescens*; *Bacillus circulans*, *Bacillus subtilis*; *Rahnella* sp.; *Acinetobacter rhizosphaerae* BIHB 723; *Bacillus subtilis* CKS1	Phosphate/iron/ nitrogen	Rong Tong, Lahaul and Spiti, India	7,13,14,15
3.	*Bacillus subtilis* CB_8A	Phosphate	Chamba, Kullu, Shimla and Kinnaur, India	16
4.	*Pseudomonas aeruginosa* strain An-15-Mg	Phosphate, iron	Shimla and Kullu, India	17
5.	*Enterobacter ludwigii* PS1; *Rhodotorula* sp. PS4; *Arthrobacter* sp.; *Cellulosimicrobium cellulans*; *Citricoccus* sp.; *Sanguibacter antarcticus*; *Sphingobacterium* sp; *Bacillus anthracis*; *Bacillus firmus*; *Bacillus flexus*; *Bacillus pumilus*; *Bacillus subtilis*; *Desemzia incerta*; *Exiguobacterium antarcticum*; *Lysinibacillus fusiformis*; *Paenibacillus terrae*; *Planococcus donghaensis*; *Klebsiella* sp.; *Providencia sp; Pseudomonas frederiksbergensis*; *Pseudomonas peli*; *Pseudomonas putida*; *Pseudomonas reactans*; *Pseudomonas stutzeri*; *Psychrobacter glacincola*; *Vibrio metschnikovii*; *Janthinobacterium* sp.	Phosphate	Leh-Ladakh, India	18,19,20
6.	*Acinetobacter*, *Pseudomonas*, *Brevibacillus*, *Gordonia terrae*, *Chryseobacterium lathyri*, *Ensifer sesbaniae*, *Paenibacillus illinoisensis*, *Penicillium oxalicum* I1 (P-I1), *Paenibacillus mucilaginosus KNP413*, *Paenibacillus mucilaginosus KNP414*	Phosphate/ Potassium	Xinping, China; Tianmu Mountain, Zhejiang, China	21,22,23
7.	*Arthrobacter sulfonivorans*; *Brachybacterium* sp.; *Cellulosimicrobium cellulans*; *Sanguibacter antarcticus*; *Bacillus firmus*; *Bacillus licheniformis*; *Bacillus muralis*; *Bacillus pumilus*; *Bacillus subtilis*; *Desemzia incerta*; *Exiguobacterium* sp.; *Lysinibacillus fusiformis* ; *Planococcus donghaensis*; *Brevundimonas terrae*; *Janthinobacterium* sp; *Klebsiella* sp.; *Pseudomonas frederiksbergensis*; *Pseudomonas putida*; *Pseudomonas reactans*; *Psychrobacter glacincola*	Iron	Leh-Ladakh, India	24

and Chrome Azurol-S for iron sequestration by the potential microbes.[30] Further characterisation of potential isolates is necessary before proceeding to the field trials using biochemicals[28,31] followed by 16S/18S rRNA gene sequencing for their identification.[31]

5.4 PHOSPHATE, POTASSIUM AND IRON SOLUBILISING MICROBES

5.4.1 Phosphate solubilising microbes and their beneficial role

Phosphorus is a growth essential and limiting nutrient for plants.[11,12] In plants, it is necessarily required for many metabolic mechanisms, like signal transduction, photosynthesis, respiration, and energy transfer. It also has importance in many other essential processes, i.e., root elongation, proliferation and seed development.[32] Therefore, it is considered as an essential nutrient that can limit the growth of the plant. Phosphorus is omnipresent, but its low solubility in soil affects its availability and makes it mostly unavailable to plants.[11,33] The phosphorus becomes unavailable in the soil due to its fixation by either adsorption on minerals/particles or precipitation by free ions[34] such as Al^{3-} and Fe^{3+}. Thus, artificial phosphorus added to soils goes into the immobile pools by reacting with aluminium and iron in acidic soil and calcium (Ca^{2+}) in calcareous soil and thus cannot be directly accessed by plants.[12]

In this context, microbes solubilising phosphate are known to play an essential role by mobilising insoluble phosphates into an available form for plants.[13] These microbes mobilise native and added phosphorus (phosphorus fertiliser) as well as the insoluble form of phosphorus; therefore, these microbes are referred to as phosphate mobilising/solubilising microbes.[12,35] In addition, the application of phosphate solubilising microbial biofertilisers to soil can improve phosphate availability to plants and thereafter, will increase crop yields.[11,33] Thus, it can be said that the presence of efficient phosphate solubilising microbes can make maximum phosphorus available to the plant for a better growth rate.[11] Further, depending on the environmental surroundings, the population and the activity of phosphate solubilising microbes vary among different regions, including the trans-Himalayan region.[11]

5.4.2 Mechanism of phosphate solubilisation

Phosphate solubilising microbes solubilise phosphorus by mineralisation, i.e., enzymatic hydrolysis facilitating the translocation of phosphate.[36] They mobilise unavailable phosphorus for plants by two processes, the direct and indirect processes. In the direct process, microbes lower the pH through proton extrusion and release of low-molecular-weight organic anions for solubilising phosphate.[37] This process of exchange of anions for phosphorus on soil adsorption sites is commonly known as *ligand exchange*. Organic acids produced by microbes, like succinic acid, having hydroxyl (OH) and carboxyl (COOH) groups, chelate the bound cations on phosphate, thereby converting phosphorus into a soluble form.[12,38] In addition, these microbes also hydrolyse organic compounds by releasing phosphatases and phytases. Thus, the insoluble phosphorus in the soil gets transformed from an inorganic, unavailable form of phosphorus into an accessible form as hydrogen phosphate (HPO_4^{2-}) and dihydrogen phosphate (H_2PO^{4-}) due to the release of organic acids, the process of chelation and ion-exchange reactions.[15] Hence, using PSM in the production of crops will decrease the cost of manufacturing expensive phosphatic fertilisers as well as making the insoluble form of phosphorus fertilisers available when inoculated in soil.[39]

Besides these processes, there are several indirect ways through which microbes can mobilise phosphorus: first by decreasing the pH, then by release of dissolved CO_2 during respiration in the soil for forming carbonic acid, and also by releasing protons (H^+) during assimilation of NH^{4+}, causing phosphorus solubilisation.[12] Phosphorus solubilising bacteria also have the capability to pull out or to absorb phosphate for re-establishing the equilibrium of phosphorus in soil through the indirect dissolution of Ca–P.[12] Providing soluble phosphorus to plants by these organisms can lead to promotion and development of plant growth by changing the concentration of phytohormones such as indoleacetic acid, N_2 fixation, siderophore production, and the potential to synthesise the enzyme ACC deaminase), which breaks down 1-aminocyclopropane-1-carboxylate into ammonium (NH_3) and $CH_3CH_2C(O)CO_2H$ (α-ketobutyrate).[40,41] This leads to the solubilisation of metal and therefore can decrease the toxicity caused by heavy metals present in stressed soils.[40]

5.4.3 Potassium solubilising bacteria and their beneficial role

Potassium (K) is the crucial macronutrient for determining soil quality and plant growth. It is greatly needed by plants as it participates in the cellular osmotic pressure adjustment and transportation of essential compounds crucial for their development and growth. Potassium leads to the activation of enzymes, nitrogen utilisation, sugar and protein synthesis, and photosynthesis in plants.[32] Moreover, it participates in vital cell metabolic mechanisms and is mostly associated with the movement of water and nutrients in plant tissues. Therefore, it is required in abundance by the plant for obtaining the highest yields.[32]

Potassium is present in trace amounts ranging from 0.04% to 3% in soil. However, despite being in a limited quantity, 98% of potassium is present in a complex with the soil phyllosilicate structure (silicate layer present in the fractions of silt and clay).[42] The remaining available percentage of potassium for plants exists either in solution form or on exchange sites.[42] This unavailability is due to the incorporation of potassium in mineral crystal lattice structures, which makes most of the soil potassium (90–98%) unavailable. Moreover, the quantity of potassium present in soil is mainly decided by its soil type, and therefore, these microbes' richness is affected by the physio-chemical characteristics of soil.

Potassium in soil usually exists in exchangeable, non-exchangeable, soluble and structural states, which are inaccessible for plants.[32] And, potassium solubilising microbes here play a fundamental role in making potassium available in soil. They also have an essential role in building humus and mineral cycling, which restricts the stocking up of dead organic materials.[43] Besides these potential benefits, these microbes also have an essential role in forming water-stable soil aggregates for improving the soil structure. In addition, these microorganisms produce essential substances causing plant growth promotion by inhibiting root pathogens.[43]

However, scanty reports are available on potassium solubilising bacteria in trans-Himalayan regions, as researchers are mostly focused on microbes having phosphate solubilising activity plus some additional activities including potassium and iron solubilising activity rather than recovering only potassium solubilising bacteria.

5.4.4 Mechanism of potassium solubilisation

Sheng and He[44] suggested that potassium solubilising bacteria like *Bacillus edaphicus* and its mutants enhance the production of citrate (citric acid), ethanedioic (oxalic) acid, dihydroxy butanedioic (tartaric) acid, butanedioic (succinic) acid and ketogluconic acid for mobilising potassium from potassium-containing minerals by chelating ions of silicon.[44] Other than bacteria, fungi can also solubilise potassium by releasing anions from organic acids (citrate, oxalate and malate), and these anions enhance the chelation of cations bound along with potassium.[45,47] In acidic soil, potassium solubilisation is found to be higher than the solubilisation of calcium and magnesium by fungi.[46] For example, *Aspergillus* strains like *A. terreus* and *A. niger* are found to have potential for boosting potassium levels by solubilising potassium-aluminium silicate ($AlKO_6Si_2$) and insoluble feldspar.[46] The major processes that cause mobilisation of potassium are acidolysis and complexolysis reactions.[47,48] During acidolysis, rhizospheric microbials chelate aluminium and silicon cations bound with potassium minerals, enhancing the exchange of potassium between soil and plants. This plays a vital part in mobilising or solubilising unavailable forms of potassium into the soil, resulting in increased potassium availability.[49]

5.4.5 Iron solubilising bacteria and their beneficial role

Iron is the most abundantly existing micronutrient in nature. It is required in the ferric form (Fe^{3+}) by all living organisms. Also, iron is considered as a fundamental factor for determining the fertility of soil. It is present in the oxidised ferrous form (Fe^{2+}), hence resulting in the formation of insoluble compounds,[35] which then become unavailable for plant or microbes in the soil. Thus, tough competition exists between plants and microbes for fulfilling the iron requirements from the rhizosphere. So, some microbes such as bacteria either form siderophores for chelating and solubilising iron from mineral or organic compounds or take up iron complexes by the cell membrane to reduce iron from Fe^{3+} to the Fe^{2+} form.[50] The unavailability of iron occurs due to bonding with heavy metals. This causes a deficiency of iron in soil, encouraging the bacterial strains to produce siderophores for the chelation of the metals from the iron–metal complex to increase the iron availability in the rhizospheric soils.[51] Hence, siderophores produced by this microbial culture have a crucial role in lowering stresses imposed on the plants by exposure to heavy metals. For example, siderophorogenic *Pseudomonas putida* and *Pseudomonas aeruginosa* strains have good iron chelating properties.

5.4.6 Mechanism of iron (Fe) solubilisation

Under the conditions of iron deficiency stress, microorganisms and plants commonly rely upon the chelating agents known as siderophores for solubilising and transporting inorganic iron, consequently increasing the bioavailability of Fe.[52] Different soil microorganisms stimulate the availability of iron by reactions including oxidation, reduction and formation/decomposition of complexes. Acidic conditions increase the solubility of iron from iron hydroxides as well as the activity of acidophiles for dissolution and bioleaching of sulphide minerals (pyrite). At neutral pH, microbes are responsible for the passive formation of Fe(III) oxides, although active mineralisation of the Fe oxides can also occur. Certain microbes oxidise iron from the Fe(II) state to the Fe(III) state and then precipitate it as ferric hydroxide($Fe(OH)_3$) around the cells. Moreover, many heterotrophic species convert soluble organic iron salts into inorganic salts, and therefore, soil bacteria and fungi produce certain acids (organic, carbonic, nitric and sulphuric acids) for reforming inorganic Fe to readily available Fe for plants.[53]

5.5 CONCLUSION AND FUTURE ASPECTS

Based on the preceding discussions, it can be concluded that the trans-Himalayan regions are found to be enriched in various PGPMs, which are proficient in the utilisation of insoluble minerals (phosphate, potassium and iron) through enzymatic activities, pH reduction, siderophore production, etc. To the best of our knowledge, it has been found that maximum research has been undertaken to isolate and understand the phosphate solubilising microbes' interaction with soil. However, scanty reports are available related to other mineral solubilising microbes, which need more investigation on their role in soil fertility and plant productivity. Meanwhile, it was found that PGPMs from trans-Himalayan regions have only been explored at the laboratory level. Therefore, in future, field research on these PGPMs and their consortia in cold climates and mountainous areas is required in order to know their benefits.

ACKNOWLEDGEMENT

All the authors would like to acknowledge the Director, DRDO-Defence Institute of High Altitude Research (DIHAR), Leh, Ladakh UT, India for providing support and facilities to complete this work.

REFERENCES

1. Dar, Javid Ahmad, and R. K. Dubey. "Desertification of trans-Himalayan glacial valleys-an indicator of climatic fluctuation and instability." In Proceedings of Environmental Implication and Management in Mining and Mineral Based Industries, IIT-BHU, India (2013): 41–50.
2. Charan, G., V. K. Bharti, S. E. Jadhav et al. "Altitudinal variations in soil physicochemical properties at cold desert high altitude." *Journal of Soil Science and Plant Nutrition* 13, no. 2 (2013): 267–277.
3. Sidhu, G. S., and Jaya N. Surya. "Soils of North-Western Himalayan eco-system and their land use, constraints, productivity potentials and future strategies." *Agropedology* 24, no. 01 (2014): 1–19.
4. Fitzsimons, Michael S., and R. Michael Miller. "The importance of soil microorganisms for maintaining diverse plant communities in tallgrass prairie." *American Journal of Botany* 97, no. 12 (2010): 1937–1943.
5. Ramakrishna, Wusirika, Radheshyam Yadav, and Kefeng Li. "Plant growth promoting bacteria in agriculture: Two sides of a coin." *Applied Soil Ecology* 138 (2019): 10–18.
6. Suyal, Deep Chandra, Divya Joshi, Saurabh Kumar et al. "Himalayan microbiomes for agro-environmental sustainability: Current perspectives and future challenges." *Microbial Ecology* 84 (2021): 643–675.
7. Sharma, Ranjna, Joginder Pal, and Mohinder Kaur. "Isolation of phosphate solubilizing Pseudomonas strains from apple rhizosphere in the Trans Himalayan region of Himachal Pradesh, India." *BioRxiv* (2017): 193672.
8. Sharma, S. P., and P. D. Sharma. "Improving soil health for sustainable development of agriculture in Himalaya." ADG (Soils), Indian Council of Agricultural Research, Krishi Anusandhan Bhavan-II, Pusa, New Delhi, p12, 2006.
9. Vurukonda, Sai Shiva Krishna Prasad, Davide Giovanardi, and Emilio Stefani. "Plant growth promoting and biocontrol activity of Streptomyces spp. as endophytes." *International Journal of Molecular Sciences* 19, no. 4 (2018): 952.
10. Yadav, Ajar Nath, Shashwati Ghosh Sachan, Priyanka Verma, and Anil Kumar Saxena. "Prospecting cold deserts of north western Himalayas for microbial diversity and plant growth promoting attributes." *Journal of Bioscience and Bioengineering* 119, no. 6 (2015): 683–693.
11. Chatli, Anshu S., Viraj Beri, and B. S. Sidhu. "Isolation and characterisation of phosphate solubilising microorganisms from the cold desert habitat of Salix alba Linn. in trans Himalayan region of Himachal Pradesh." *Indian Journal of Microbiology* 48, no. 2 (2008): 267–273.
12. Rashid, Muhammad Imtiaz, Liyakat Hamid Mujawar, Tanvir Shahzad et al. "Bacteria and fungi can contribute to nutrients bioavailability and aggregate formation in degraded soils." *Microbiological Research* 183 (2016): 26–41.
13. Wang, Junru, Fang-Jie Zhao, Andrew A. Meharg et al. "Mechanisms of arsenic hyperaccumulation in *Pteris vittata*. Uptake kinetics, interactions with phosphate, and arsenic speciation." *Plant Physiology* 130, no. 3 (2002): 1552–1561.
14. Mehta, Phosphate, A. Walia, A. Chauhan et al. "Phosphate solubilisation and plant growth promoting potential by stress tolerant *Bacillus sp.* isolated from rhizosphere of apple orchards in trans Himalayan region of Himachal Pradesh." *Annals of Applied Biology* 163, no. 3 (2013): 430–443.
15. Kumar, Anil, Shiwani Guleria, Preeti Mehta, Abhishek Walia, Anjali Chauhan, and Chand Karan Shirkot. "Plant growth-promoting traits of phosphate solubilizing bacteria isolated from *Hippophae rhamnoides* L. (Sea-buckthorn) growing in

cold desert Trans-Himalayan Lahul and Spiti regions of India." *Acta Physiologiae Plantarum* 37, no. 3 (2015): 1–12.
16. Gulati, Arvind, Praveen Rahi, and Pratibha Vyas. "Characterization of phosphate-solubilizing fluorescent pseudomonads from the rhizosphere of seabuckthorn growing in the cold deserts of Himalayas." *Current Microbiology* 56, no. 1 (2008): 73–79.
17. Gulati, Arvind, Pratibha Vyas, Praveen Rahi, and Ramesh Chand Kasana. "Plant growth-promoting and rhizosphere-competent Acinetobacter rhizosphaerae strain BIHB 723 from the cold deserts of the Himalayas." *Current Microbiology* 58, no. 4 (2009): 371–377.
18. Vyas, Pratibha, Robin Joshi, K. C. Sharma, Praveen Rahi, Ashu Gulati, and Arvind Gulati. "Cold-adapted and rhizosphere-competent strain of Rahnella sp. with broad-spectrum plant growth-promotion potential." *Journal of Microbiology and Biotechnology* 20, no. 12 (2010): 1724–1734.
19. Mundra, Sunil, Richa Arora, and Tsering Stobdan. "Solubilization of insoluble inorganic phosphates by a novel temperature-, pH-, and salt-tolerant yeast, *Rhodotorula* sp. PS4, isolated from seabuckthorn rhizosphere, growing in cold desert of Ladakh, India." *World Journal of Microbiology and Biotechnology* 27, no. 10 (2011): 2387–2396.
20. Dolkar, Diskit, Phuntsog Dolkar, Stanzin Angmo et al. "Stress tolerance and plant growth promotion potential of *Enterobacter ludwigii* PS1 isolated from Seabuckthorn rhizosphere." *Biocatalysis and Agricultural Biotechnology* 14 (2018): 438–443.
21. Gong, Mingbo, Peng Du, Xue Liu, and Changxiong Zhu. "Transformation of inorganic P fractions of soil and plant growth promotion by phosphate-solubilizing ability of *Penicillium oxalicum* I1." *Journal of Microbiology* 52, no. 12 (2014): 1012–1019.
22. Jiang, Huanhuan, Tong Wang, Xiaoyuan Chi et al. "Isolation and characterization of halotolerant phosphate solubilizing bacteria naturally colonizing the peanut rhizosphere in salt-affected soil." *Geomicrobiology Journal* 37, no. 2 (2020): 110–118.
23. Hu, Xiufang, Jishuang Chen, and Jiangfeng Guo. "Two phosphate-and potassium-solubilizing bacteria isolated from Tianmu Mountain, Zhejiang, China." *World Journal of Microbiology and Biotechnology* 22, no. 9 (2006): 983–990.
24. Yadav, Janardan, Jay Prakash Verma, Durgesh Kumar Jaiswal, and Ashok Kumar. "Evaluation of PGPR and different concentration of phosphorus level on plant growth, yield and nutrient content of rice (*Oryza sativa*)." *Ecological Engineering* 62 (2014): 123–128.
25. Barillot, Cindy DC, Claude-Olivier Sarde, Valerie Bert et al. "A standardized method for the sampling of rhizosphere and rhizoplan soil bacteria associated to a herbaceous root system." *Annals of Microbiology* 63, no. 2 (2013): 471–476.
26. Kiran, Arigela, and P. P. Kadu. "Effect of seed inoculation of zinc and iron solubilizing microorganisms on soil microbial count as influenced by different treatments at panicle initiation and harvest stage of wheat in inceptisol." *International Journal of Current Microbiology and Applied Sciences* 8, no. 4 (2019): 814–821.
27. Aneja, K. R. Experiments in microbiology, plant pathology and biotechnology. Fourth edition. pp. 1–604, *New Age International* Limited Publishers, New Delhi, 2007.
28. Mwajita, Mwashasha Rashid, Hunja Murage, Akio Tani, and Esther M. Kahangi. "Evaluation of rhizosphere, rhizoplane and phyllosphere bacteria and fungi isolated from rice in Kenya for plant growth promoters." *SpringerPlus* 2, no. 1 (2013): 1–9.

29. Sharon, J. A., L. T. Hathwaik, G. M. Glenn et al. "Isolation of efficient phosphate solubilizing bacteria capable of enhancing tomato plant growth." *Journal of Soil Science and Plant Nutrition* 16, no. 2 (2016): 525–536.
30. Batista, FC, CS Abreu, VP Ribeiro et al. "Iron phosphate solubilization and siderophore production by microorganisms isolated from maize." In Embrapa Milho e Sorgo-Resumo em anais de congresso (ALICE) (ed) Congresso Brasileiro De Microbiologia, 29.; Encontro Nacional De Microbiologia Ambiental, 15, 2017.
31. Pande, Amit, Prashant Pandey, Simmi Mehra, Mritunjay Singh, and Suresh Kaushik. "Phenotypic and genotypic characterization of phosphate solubilizing bacteria and their efficiency on the growth of maize." *Journal of Genetic Engineering and Biotechnology* 15, no. 2 (2017): 379–391.
32. Kumawat, Narendra, Rakesh Kumar, Sudhir Kumar, and Vijay Singh Meena. "Nutrient solubilizing microbes (NSMs): Its role in sustainable crop production." In Agriculturally Important Microbes for Sustainable Agriculture, pp. 25–61. Springer, 2017.
33. Thomas, Lebin, and Ishwar Singh. "Microbial biofertilizers: Types and applications." In *Biofertilizers for Sustainable Agriculture and Environment*, pp. 1–19. Springer, 2019.
34. Sharma, Seema B., Riyaz Z. Sayyed, Mrugesh H. Trivedi, and Thivakaran A. Gobi. "Phosphate solubilizing microbes: Sustainable approach for managing phosphorus deficiency in agricultural soils." *SpringerPlus* 2, no. 1 (2013): 1–14.
35. Ma, Jing, Erdeng Ma, Hua Xu et al. "Wheat straw management affects CH4 and N2O emissions from rice fields." *Soil Biology and Biochemistry* 41, no. 5 (2009): 1022–1028.
36. Owen, D., A. P. Williams, G. W. Griffith, and P. J. A. Withers. "Use of commercial bio-inoculants to increase agricultural production through improved phosphrous acquisition." *Applied Soil Ecology* 86 (2015): 41–54.
37. Chen, Y. P., P. D. Rekha, A. B. Arun et al. "Phosphate solubilizing bacteria from subtropical soil and their tricalcium phosphate solubilizing abilities." *Applied Soil Ecology* 34, no. 1 (2006): 33–41.
38. Zhang, Lin, Jiequn Fan, Xiaodong Ding et al. "Hyphosphere interactions between an arbuscular mycorrhizal fungus and a phosphate solubilizing bacterium promote phytate mineralization in soil." *Soil Biology and Biochemistry* 74 (2014): 177–183.
39. Kumar, Rakesh, and Narendra Kumawat. "Effect of sowing dates, seed rates and integrated nutrition on productivity, profitability and nutrient uptake of summer mungbean in Eastern Himalaya." *Archives of Agronomy and Soil Science* 60, no. 9 (2014): 1207–1227.
40. Zaidi, A., M. Khan, Munees Ahemad, and M. Oves. "Plant growth promotion by phosphate solubilizing bacteria." *Acta microbiologica et immunologica Hungarica* 56, no. 3 (2009): 263–284.
41. Wani, Parvaze A., Mohammad Saghir Khan, and Almas Zaidi. "Synergistic effects of the inoculation with nitrogen-fixing and phosphate-solubilizing rhizobacteria on the performance of field-grown chickpea." *Journal of Plant Nutrition and Soil Science* 170, no. 2 (2007): 283–287.
42. Shelobolina, Evgenya, Eric Roden, Jason Benzine, and Mai Yia Xiong. "Using phyllosilicate-fe (ii)-oxidizing soil bacteria to improve fe and k plant nutrition." U.S. Patent Application 14/209, 509, filed September 18, 2014.
43. Zakaria, Ahmad Azinuddin Bin. "Growth optimization of potassium solubilizing bacteria isolated from biofertilizer." *Faculty of Chemical & Natural Resources Engineering*, University Malaysia Pahang, 2009.

44. Sheng, Xia Fang, and Lin Yan He. "Solubilization of potassium-bearing minerals by a wild-type strain of Bacillus edaphicus and its mutants and increased potassium uptake by wheat." *Canadian Journal of Microbiology* 52, no. 1 (2006): 66–72.
45. Meena, Vijay Singh, B. R. Maurya, and Jay Prakash Verma. "Does a rhizospheric microorganism enhance K+ availability in agricultural soils?" *Microbiological Research* 169, no. 5–6 (2014): 337–347.
46. Prajapati, Kalavati, M. C. Sharma, and H. A. Modi. "Isolation of two potassium solubilizing fungi from ceramic industry soils." *Life Sciences Leaflets* 5 (2012): 71–75.
47. Uroz, Stéphane, Christophe Calvaruso, Marie-Pierre Turpault, and Pascale Frey-Klett. "Mineral weathering by bacteria: Ecology, actors and mechanisms." *Trends in Microbiology* 17, no. 8 (2009): 378–387.
48. Parmar, Priyanka, and S. S. Sindhu. "Potassium solubilization by rhizosphere bacteria: Influence of nutritional and environmental conditions." *Journal of Microbiology Research* 3, no. 1 (2013): 25–31.
49. Abou-el-Seoud, Islam I., and Ahmed Abdel-Megeed. "Impact of rock materials and biofertilizations on P and K availability for maize (*Zea Maize*) under calcareous soil conditions." *Saudi Journal of Biological Sciences* 19, no. 1 (2012): 55–63.
50. Machuca, A., G. Pereira, A. Aguiar, and A. M. F. Milagres. "Metal-chelating compounds produced by ectomycorrhizal fungi collected from pine plantations." *Letters in Applied Microbiology* 44, no. 1 (2007): 7–12.
51. Miethke, Marcus, and Mohamed A. Marahiel. "Siderophore-based iron acquisition and pathogen control." *Microbiology and Molecular Biology Reviews* 71, no. 3 (2007): 413–451.
52. Kong, Zhaoyu, and Bernard R. Glick. "The role of plant growth-promoting bacteria in metal phytoremediation." *Advances in Microbial Physiology* 71 (2017): 97–132.
53. Chikkanna, Arpitha, and Devanita Ghosh. "Microbial mineral dissolution and environmental disasters: Microbes and their mineral interactions." In *Microbial Biotechnology in Environmental Monitoring and Cleanup*, pp. 125–151. IGI Global, 2018.

6 Potential and responses of microfungi to the dynamics of cold environments

Skarma Nonzom and Geeta Sumbali*

CONTENTS

6.1 INTRODUCTION

Cold habitats may vary from high mountains to deep sea and from the Arctic to the Antarctic. It is estimated that more than 80% of Earth is cold, of which approximately 71% is occupied by oceans (1 °C to 4 °C), 35% by snow cover, 24 % by frozen ground, 13% by sea ice and 10% by glaciers.[1] Other cold environments are represented by cold deserts and soils, caves, sand and lakes. The organisms that are adapted and survive in such cold habitats have been initially grouped as psychrophiles and psychrotrophs. The former forms are constrained to permanently cold environments, like the polar regions, lakes, oceans, ice and snow fields, caves and alpine soils, as they are specifically adapted to a low temperature of 0 °C or below, with optimum temperatures that range up to15 °C and a maximum temperature of around 20 °C. Psychrotolerant forms, in contrast, have

* Corresponding author e-mail: skarmanunzom@yahoo.com

DOI: 10.1201/9781003354031-6

a wider range of habitats, with optimum growth temperatures of up to 20–25 °C and maximum temperatures of as high as 40 °C.[2]

6.2 MICROFUNGAL DIVERSITY FROM DIFFERENT SOILS

Soil is a highly complex medium. Diverse groups of microorganisms are known to occur in the soil. Microfungi represent one of the dominant groups, characteristically constituting more biomass than the bacteria, depending on the growth and nutrient conditions. They are known to survive in most habitats, surpassing bacteria due to their capability to withstand extreme environmental conditions.[3]

There have been ample reports on soil microbes from diverse areas of the globe since the beginning of studies on soil microbes (Table 6.1). Literature reports greater prevalence and high frequency of genera like *Penicillium*, *Aspergillus*, *Trichoderma* and *Fusarium* in various ecosystems.[4] The high prevalence and dominance of these genera in different soils of various geographical areas could be due to their highly sporulating and rapid growth ability.

Also, these microfungal genera possess a broad ecological spectrum.[5] Of the various microfungi, the predominance of *Penicillium* in soils of temperate and cold forests has also been documented by many researchers.[4,6] For example, Reddy[7] reported *Penicillium* as one of the dominant fungi in Nilgiri high-altitude forest soils. Recently, the dominance of Penicilli, especially with monoverticillate branching, has been well documented in the world's second-coldest inhabited place (Drass).[8] Cold-adapted Penicilli have also been documented by various researchers in Antarctica – for example, from lakes,[9] historical woodlands,[10] soils,[11] algal thalli[12] and so on. However, Aspergilli are usually found inhabiting tropical soils with a warm climate, showing lower frequency in temperate soils.[6] According to Christensen,[13] dematiaceous microfungi constitute a specific group of mycoflora of soil. Further, there are reports that species of genera such as *Alternaria*, *Cladosporium*, *Aspergillus*, *Stachybotrys*, *Ulocladium* and *Dreshlera* represent the most frequent and abundant microfungi.[14,15]

However, at present, a range of studies (Table 6.2) reveals that psychrophiles could successfully inhabit and survive in the non-Antarctic cryosphere, such as Arctic regions and glaciers of high mountains like the Alps, the Andes and the Himalayas.[16]

The occurrence of psychrophilic and psychrotolerant/psychrotrophic microfungi has been well documented in cold habitats, such as cold water,[45] permafrost regions,[46] ice masses, glaciers, freshwater ice, sea ice as well as icebergs,[47] along with coastal waters of polar regions.[48] Studies carried out by Blanchette and coworkers[49] on Ross Island showed the association of endemic microfungi with the historic expedition huts. Similarly, the dry polar soils have been reported to harbour only yeasts as endemic mycoflora.[50] It has been reported that yeasts predominate in the undisturbed regions of the dry valleys of Antarctica. Earlier, Cameron and coworkers (1976)[51] observed *Cryptococcus albidus* to be the most frequent cold-tolerant species. The diversity of microfungi recovered from dry regions of Antarctica was found to be similar to that of dry Arctic soils.[52] Several

TABLE 6.1
Reports of microfungal diversity from cold environments

Source of the microfungi	Cold environment/collection site	Dominant taxa	Reference
Soil	Mussoorie	*Chrysosporium keratinophilum*	17
NM	Uttarakhand	*C. merdarium*	18
Alpine plant *Ranunculus adoneus*	Niwot Ridge Long Term Ecological Research (LTER) site in the Colorado Front Range	*Phialocephala fortinii* (dark-septate endophyte)	19
Glacier bank soils	Kashmir valley	*Microsporum gypseum*	20
Soil	Shimla, Manali, Mandi, Chamba and Dharamshala of Himachal Pradesh	*Chrysosporium queenslandicum*	21
Ectomycorrhizal on *Salix herbacea*	Glacier forefronts in the Tyrolean Alps	*Cenococcum geophilum*, *Sebacina* sp., *Tomentella* sp., *Cortinarius* sp.	22
Periglacial soils	Annapurna Conservation Area (ACA) of the Himalaya Mountains (Nepal) and the Front Range of the Colorado Rocky Mountains (USA)	*Batrachochytrium dendrobatidis*, *Spizellomyces* sp.	23
Air	Historic huts on Ross Island	*Antarctomyces psychrotrophicus*, *Cladosporium cladosporioides*, *Geomyces* sp., *Pseudeurotium desertorum*	24
Soil	Cold desert of Ladakh	*Chrysosporium tropicum*	25
Ectomycorrhizal fungi associated with *Salix polaris* Wahlenb	Austre Broggerbreen, Svalbard	*Geopora* sp. and *Cenococcum* sp.	26
Cold-adapted endophytic fungi	Baima Snow Mountain (altitude 4000–4300 m), south-west China	*Cephalosporium*, *Sirococcus*, *Penicillium*, *Aspergillus*	27

(*Continued*)

TABLE 6.1 (CONTINUED)
Reports of microfungal diversity from cold environments

Source of the microfungi	Cold environment/collection site	Dominant taxa	Reference
Soil (vegetational and climatic zones)	Eastern Himalayan Range	*Penicillium*, *Aspergillus*	28
Soil from potato fields	Mana village, Chamoli district, Uttarakhand	*Penicillium*	29
Cold arid base soil	Moonland landscape (Ladakh)	*Alternaria alternata*, *Aspergillus flavus*, *A. fumigatus*, *A. ochraceus*, *A. sydowii*, *A. niger*, *Cladosporium cladosporiodes*, *C. sphaerospermum*, *Penicillium griseofulvum*, *P. olivicolor*, *P. italicum*, *P. verrucosum*, *Ulocladium atrum*	15
Sediment, ice water	Batura Passu and Siachen Glaciers	*Penicillium*, *Cladosporium*, *Geomyces*, *Cordyceps*, *Mrakia*, *Cadophora*, *Tetracladium*, *Eupenicillium*, *Trametes*, *Mortierella*, *Scopulariopsis*, *Beauveria*, *Candida*, *Pseudogymnoascus*	30
Soil	Temperate mountain permafrost	*Ascomycota*	31
Bird's feathers	Spitsbergen island of Svalbard archipelago	Class *Ascomycota*	32
Rock inhabiting lichen thalli	Koralpe mountain range in the south-eastern rim of the Austrian Alps	*Chaetothyriomycetes*, *Dothideomycetes*	33
Glaciers	Qinghai–Tibet Plateau	*Phoma sclerotioides*, *Pseudogymnoascus pannorum*	34
Siachen glacier	Himalaya Range, Pakistan	*Leotiomycetes* sp.	35
Soil	Swiss target region of Gloria, Switzerland	*Chloroflexi* and Mucormycota	36

(Continued)

TABLE 6.1 (CONTINUED)
Reports of microfungal diversity from cold environments

Source of the microfungi	Cold environment/collection site	Dominant taxa	Reference
Alpine wetlands	Huahu Lake wetland of Zoige national reserve wetland, Mangcuo Lake wetland, Baima Snow Mountain nature reserve wetland and Hengduan Mountains of Qinghai–Tibet Plateau	Class Ascomycota	37
Soil	Colombia	Phylum Glomeromycota	38
Soil	Drass valley, Kargil	*Penicillium adametzi*, *P. citrinum*, *P. herquei*, *P. implicatum*, *P. italicum*, *P. resedanum*, *P. restrictum*, *P. roseopurpureum*, *P. sclerotiorum*, *P. verrucosum*, *P. vinaceum*	8
Leaf and root endophyte of *Arnebia euchroma*	Langza and Spiti of Himachal Pradesh	*Eurotiales* and *Penicillium*	39
Freshwater pond	Arctic	*Leotiomycetes* and *Dothideomycetes*	40
Sediment and soil	Brestovská Cave and Demänovská L'adová and Slobody Caves (Slovakia)	*Arthroderma quadrifidum*	41
Cave ecosystem	Castañar Cave (south-west Spain)	*Candida parapsilosis*, *Sistotrema oblongisporum*, *Cephalotrichum microsporum*, *Mortierella alpina*, *Omphalotus olearius*, *Trichophyton ajelloi*, *Stereum hirsutum*, *Malassezia globose*	42
Soil	Alpine–gorge region on the eastern Qinghai–Tibetan Plateau	Phylum Ascomycota	43
Castañar Cave	Spain	*Candida parapsilosis*	42
Coal mine spoil tips	Susuman (Magadan Region, Russia)	*Pseudogymnoascus pannorum* and *Penicillium*	44

TABLE 6.2
Reports of microfungal diversity from the polar cold environments

Source of the microfungi	Polar cold environment/collection site	Dominant taxa	Reference
Soil	Antarctica	Genus *Chrysosporium Pseudogymnoascus* sp.	56
Soil from the coast of Ross Sea and dust samples inside the Italian scientific base	Antarctic	NM	57
Arctic soil	Franz Joseph Land	*Arachniotus citrinus* and *Phoma* sp., *Pseudogymnoascus pannorum*	52
Bird excreta, skeletal remains	Schirmacher Oasis, East Antarctica	*Phoma herbarum*	58
Lichens		*Acremonium antarcticum* and *A. psychroph*	
Ornithogenic soil		*Torulopsis* sp., *Psychrophila* sp., *Cryptococcus* sp.	
Oil spills		*Hormoconis resinae*	
Soil	South Shetland Islands, Antarctica	*Antarctomyces psychrotrophicus*	59
Mosses	Antarctica	*Pseudogymnoascus* sp.	60
Carex aquatilis, C. paleacea, C. paleacea × aquatilis, C. recta, C. acuta	Northern Fennoscandia, Arctic	*Anthracoidea echinospora, A. heterospora*	61
Sediment-rich subglacial ice	Arctic glaciers (Svalbard, Norway)	*Penicillium crustosum*	62
Thalli of macroalgae	Antarctica	*Pseudogymnoascus* sp.	12
Leaves of *Colobanthus quitensis*	Antarctica	*Pseudogymnoascus* sp.	53
Soil	McMurdo Dry Valleys of Antarctica	*Zalerion varium*	63
Lakes	Antarctica	*Geomyces pannorum, Thelebolus* sp., *Mortierella* sp.	64
Freshwater lakes		*Pseudogymnoascus* sp.	
Soil	Hornsund, Spitsbergen in the High Arctic	*Phialocephala lagerbergii, Mortierella macrocystis*	65

(*Continued*)

TABLE 6.2 (CONTINUED)
Reports of microfungal diversity from the polar cold environments

Source of the microfungi	Polar cold environment/collection site	Dominant taxa	Reference
Endophytic fungi of *Barbilophozia hatcheri*	King George Island, maritime Antarctica	*Hyaloscyphaceae* sp.	66
Chorisodontium aciphyllum, S. uncinata		*Rhizoscyphus* sp.	
Lichens	Antarctica	*Pseudogymnoascus* sp.	67
Endophytic fungi associated with *Cassiope tetragona, Saxifraga cespitosa, S. oppositifolia* and *Silene acaulis*	Ny-Ålesund region, Spitsbergen, the largest island of the Svalbard archipelago (High Arctic)	*Cryptococcus, Rhizosphaera, Mycopappus, Melampsora, Tetracladium, Phaeosphaeria, Mrakia, Venturia, Leptosphaeria*	68
Snow	King George Island, Antarctica	*Antarctomyces pellizariae*	69
Feathers from seabirds and sea ducks	Canadian Arctic	*Penicillium antarcticum*	70
Subglacial ice	Kingsford, Svalbard	*Glaciozyma*-related taxon, *Phenoliferia psychrophenolica*	71
Soil	Deception Island, South Shetland Islands, Antarctica	*Pseudogymnoascus roseus, P. appendiculatus, P.* sp., *Leotiomycetes* sp., *Penicillium* sp., *Mortirella* sp.	72
Soil	Maritime Antarctic	*Pseudogymnoascus roseus, Mortierella* sp. and *Cladosporium halotolerans*	73
Lake	Maritime Antarctica	*Tetracladium* sp.	74
Marine sediments	Antarctica	*Pseudocercosporella, Toxicocladosporium, Trichoderma, Humicola, Paraconiothyrium, Phaeoacremonium, Phenoliferi*	75

studies from diverse locations of Antarctica (Table 6.2) showed a high incidence of *Geomyces pannorum*,[15,53] which is considered to be keratinophilic[54] as well as indigenous.[55]

Based on studies carried out on soil microflora of the inland McKelvey Valley, it has been suggested that free-living microfungi are less resistant to extreme environments in comparison to bacteria, as the researchers could recover only seven fungal phylotypes, and that dispersal limitations could be the reason for low fungal incidence.[63] However, in Antarctica's coastal regions, based on aerobiological studies of bacterial and fungal spore dispersal, it has been implied that viable spores and cells are dispersed in considerable numbers and significantly over a very long distance.[24]

Mycorrhizal fungal communities have also been reported to be common in the Arctic habitats and are considered essential for the growth and survival of the host plants by providing water and required nutrients.[76] An example is the frequent distribution of Ectomycorrhizal fungal groups from the Arctics as well as the alpine habitats of almost all the continents. *Cortinarius*, *Inocybe*, *Hebeloma*, *Thelephora*, *Russula*, *Cenococcum*, *Tomentella* and *Laccaria* represent some commonly distributed Ectomycorrhizal fungal genera.[26,77]

Similarly, Zhang and Yao[68] reported *Rhizosphaera*, *Cryptococcus*, *Mycopappus*, *Phaeosphaeria*, *Tetracladium*, *Venturia*, *Mrakia*, *Leptosphaeria* and *Melampsora* as the dominant genera while assessing the diversity of fungal endophytes in the High Arctics.

In addition, some fungal plant pathogens have also been reported from the Arctic habitat, which include some obligate pathogens, such as *Exobasidium*, rusts and smuts.[61]

6.3 MICROFUNGAL DIVERSITY FROM HIMALAYAN COLD ENVIRONMENTS

The Himalayas, which are known to have the highest altitude (approximately 8000 msl), are characterised by environmental conditions similar to the other regions on the globe with extremely cold temperatures.[78] Hindu Kush–Karakoram Himalaya encompasses around 20,000 glaciers, of which 5000 fall in the Karakoram range, whereas around 12,000 are in the Himalayas, spread over an area of 60,000 km.[79] Studies on the fungal diversity (psychrotrophic or psychrophilic) associated with these glaciers are still meagre and fragmentary. In India, some reports support the presence of aquatic *Lemonniera* and *Tetracladium* of the group Hyphomycetes from the Kumaun Himalaya.[80] Deshmukh and Sunil (2002)[20] reported keratinophilic microfungi from the banks of glaciers from Sonamarg, Gulmarg, Tangmarg and Khilanmarg (Kashmir). He observed that 67.3% of the recovered microfungi, *Microsporum van-breuseghemii* (0.9%), *Malbranchea* spp. (0.9%), *M. nanum* (1.9%), *Uncinocarpus reesii* (1.9%), *Trichophyton terrestre* (2.8%), *Geomyces pannorum* (2.8%), *Chrysosporium keratinophilum* (3.7%), *C. tropicum* (5.6%), *Chrysosporium* state of *Ctenomyces serratus* (11.2%), *T. ajelloi* (15%) and *M. gypseum* (20.6%), showed the ability to utilise keratin and hence were keratinophilic. Similar observations were made by Deshmukh and Sunil (2006)[21] from soils of hilly regions of Himachal Pradesh, wherein 83% of soil samples showed the presence of keratinophilic microfungi and related dermatophytes. Lone and coworkers (2011)[81] isolated rhizospheric microfungi associated with the root of *Juglans regia* L. in the Kashmir valley. Similarly, Sagar and coworkers (2011)[82] recovered 45 microfungal taxa from the rhizosphere plants inhabiting the cold arid regions of Himachal Pradesh, India. Of these, 21 species were found to be psychrophilic, 7 were thermophilic and the remaining 17 were psychrotolerant in nature.

Later, Shivaji and coworkers[83] identified 25 psychrophilic yeasts from Roopkund Lake soil in the Himalayas, India. In a similar study, about 57 fungal taxa, with the genus *Alternaria* as the dominant one, have been reported from Tirich Mir glacier, Pakistan.[84] Wang et al.,[85] while studying the diversity of the cold-tolerant microfungi of the glaciers of Qinghai–Tibet, reported six novel species: *Psychrophila lutea*, *P. olivacea*, *P. antarctica*, *Tetracladium globosum*, *T. psychrophilum* and *T. ellipsoideum*. Similarly, the Karakoram and Hindu Kush Mountain ranges, Pakistan (Batura, Passu and Siachen glaciers) have also been investigated for the diversity of microfungi[30] with a representation of 77 fungal taxa belonging to 24 fungal genera. The majority of the recovered microfungi were represented by *Penicillium*, *Beauveria*, *Cladosporium*, *Leotiomycetes*, *Geomyces*, *Mrakia*, *Trichoderma*, *Thelebolus*, *Pseudogymnoascus*, *Fontanospora*, *Pseudeurotium*, *Periconia*, *Cordyceps*, *Cryptococcus*, *Cadophora*, *Mortierella*, *Trametes*, *Scopulariopsis*, *Candida*, *Eupenicillium*, *Phoma* and *Sporobolomyces*. *Penicillium* species showing significant tolerance to low temperature, high pH and salt concentrations have been reported from the Indian Himalayas as well.[86]

A few mycological studies on cold-tolerant fungi have also been carried out in the cold deserts of Ladakh. For example, Deshmukh et al.[25] isolated and identified 14 species of keratinophilic soil microfungi from the Ladakh region. Similar studies were reported by Kotwal and Sumbali (2011),[87] Anupama and her coworkers (2011),[88] and Nonzom and Sumbali (2015).[15] In a recent study, a total of 102 psychrotolerant microfungi belonging to 38 genera were reported from some unexplored regions of the trans-Himalaya, with 4 new records from India.[8,89] In addition to the mycological studies, bacterial diversity was also reported from these regions.[90,91]

6.4 CHALLENGES FOR SURVIVAL IN COLD ENVIRONMENTS

Cold environments present several challenges in the form of extremely low temperatures, repeated freeze and thaw cycles, high UV radiation, increased salinity, reduced moisture, desiccation, poor nutrient availability, etc. for the growth and proliferation of various existing microfungi.[92] Low temperature is known to affect fungal cells by modifying the viscosity of water, chemical reactions and membrane stability, and denaturing proteins.[93] Low temperatures also adversely affect the physical characteristics and membrane functioning, resulting in a decline of membrane fluidity leading to a gel-phase transition and ultimately, function loss.[94]

6.5 STRATEGIES FOR SURVIVAL IN COLD ENVIRONMENTS

Studies have established that specialised microbes have shown their ability to tolerate a variety of stress factors, such as hypersalinity, desiccation, high UV radiation, low temperature and so on, by developing diverse functional strategies, for example, by producing several bioactive compounds or antifreeze proteins or/and cold-active enzymes.[95] One such strategy was revealed in 2017 by Pacelli[96] and his coworkers, who characterised the effects of space-relevant and of radiation on *Cryomyces antarcticus* and observed that this microfungus was able to survive and maintain high metabolic activity without any perceptible damage to DNA even after exposure to the maximum radiation dose. As a result, researchers have been using *Cryomyces* as a model for experiments related to space, and it is considered the best (eukaryotic) model for astrobiological studies so far.[97]

The present scenario of the diversity, distribution and ecology of microbes prevailing in the cryosphere of Earth has revealed that even extreme environments with freezing temperatures harbour an exceptionally viable, diverse and metabolically active group of microbes representing some of the major phylogenetic groups.[1] In spite of the various metabolic capabilities and dynamics in the environment, one of the key features of all the cold-adapted microbes is the successful surmounting of the adverse effects of low temperatures, which they achieve by developing an array of structural and functional adaptations:[98]

1. Increased membrane fluidity has been observed in a cold-tolerant yeast *Rhodosporidium obovatum*, signifying the role of unsaturated fatty acids in maintaining plasma membrane integrity at lower temperatures.[99]
2. Trehalose, usually found in the vegetative and reproductive stages of a fungus, has also been found to play a vital role in enhancing the resistance of microfungi to various environmental stresses, such as dehydration, desiccation and freezing along with extreme temperatures.[100] It has now been established that trehalose concentration increases in microfungi upon exposure to low temperatures, as exemplified by the changes observed in a) roots of alpine mycorrhizae,[101] b) psychrophilic microfungi, *Humicola marvinii*[102] and c) *Hebeloma* species.[103]
3. Antifreeze proteins (AFPs) have been well studied in pathogenic moulds,[104] which are represented by two important families (Basidiomycetes and Ascomycetes) and one species of Oomycetes. Among the reported taxa, these proteins have solely been reported in a Basidiomycetes member, *Coprinus psychromorbidus*. Later, Hoshino[104] studied seven strains belonging to ascomycetes from Antarctica with the ability to alter ice crystal nature, though not recognised as AFPs. In another study, a new fungal AFP from *Antarctomyces psychrotrophicus*, an ascomycete of Antarctica, has been reported.[105]
4. The microfungi and other microbes that grow and survive at extremely low temperatures are well supported by the production of cold-active enzymes, since they provide active sites for the substrates, show efficient

flexibility and increase their specific activities at low energy cost.[102] A large number of microorganisms are involved in producing enzymes and many other secondary metabolites.[106] For example, in the Arctic region, an array of enzymes (keratinase, cellulase, amylase, protease, xylanase, pectinase and chitinase enzymes) have been reported, with most of the secondary metabolites reported from bacteria.[2,107,108] In other microbial groups, such as fungi,[107,108] actinomycetes,[109] microalgae[109] and cyanobacteria,[110] only a few such studies are reported in the literature. The fungal strains of Antarctica, which produce cold-active enzymes, suggest the resistance of the fungi to survive and flourish at extremely low temperatures, while their extensive enzymatic capabilities strengthen their probability of perseverance in such hostile environments.[111]

6.6 IMPORTANCE OF COLD-TOLERANT MICROFUNGI

Since the breakthrough of various vital bioactive natural products, like lovastatin and penicillin, the potential biosynthetic activities of culturable filamentous microfungi have influenced human life.[112] Earlier, one such study showed the antimicrobial (antibacterial and antifungal) properties of an Antarctic fungus, *Geomyces*.[113] Later, the ability of *Orthodendron truncatum* isolated from Antarctica (King George Island) to produce cytotoxic metabolites against various cancer lines in humans along with exhibiting potent antifungal and antimycobacterial activity was demonstrated.[114] Similar observations of cytotoxic activities have also been exhibited by fungal species like *Cladosporium sphaerospermum*, *O. truncatum* and *Mortierella* sp., which were associated with Antarctic soils.[112] Their study also established the importance of species of *Penicillium* as a producer of bioactive metabolites, thereby corroborating the results of previous similar studies by various researchers.[115] Similarly, Godinho and his coworkers[116] showed the capability of several culturable microfungi isolated from the soil of Antarctica, such as *Aspergillus sydowii*, *Penicillium brevicompactum*, *P. allii-sativi*, *P. rubens* and *P. chrysogenum*, to have various bioactivities, such as antiviral, antifungal, antibacterial, antiprotozoal, antitumoral and herbicidal. In a similar recent investigation, the bioactivity of a range of microfungi including *Penicillium* has been demonstrated, with the majority of them showing herbicidal, trypanocidal and leishmanicidal activities.[117] Similarly, the ability of *Penicillium palitans* recovered from an undisturbed 30,000-year-old permafrost (Antarctica) to produce festuclavine and fumigaclavines A and B has been recently documented.[118] Therefore, these cold-adapted microfungi are becoming vital sources for various novel bioactive secondary metabolites.

Psychrophilic and psychrotrophic microbes and their cold-adapted metabolites (enzymes/proteins) possess a multitude of applications in diverse fields.[119] These microbes become adapted to the extreme niches, resulting in the evolution of their molecular machinery, and may produce cold-adapted enzymes,[120] which are active at moderate and low temperatures[121] and thus are cost-effective.[2] Among various enzymes, α-amylases,[122] lipases,[123] proteases,[124] xylanases,[125] cellulases,[126] chitinases[127] and pectinases[128] constitute important groups that show significant catalytic efficiency at a lower temperature. The majority of the cold-active enzymes that have been studied and characterised so far involved bacteria, while studies on fungal cold-active enzymes are scarce. Recently, Nonzom and Sumbali[129] screened around 102 psychrotolerant microfungi inhabiting some cold arid soils of Kargil for their capability to produce three cold-active extracellular enzymes: cellulase, protease and lipase. Out of the total 102 recovered microfungi, which were screened for enzymatic activity, only 28% of them showed cellulase activity, 46% showed lipase activity and 58% showed protease activity. *Penicillium citrinum* and *P. italicum* showed the highest cellulase activity, whereas *Aspergillus terreus* and *Penicillium fellutinum* exhibited the lowest cellulase activity.

Some cold-active enzymes, such as chitinase, xylanase and protease, especially of *Trichoderma* and *Penicillium* species, have been documented from Antarctica. They show maximum activity at 10 °C, indicating their adaptation to the colder temperatures of that environment.[130] The psychrophilic microfungus *Geomyces pannorum* has been reported to hydrolyse starch and produce extracellular enzymes, such as lipase, chitinase and urease, that show activity even below 25 °C.[131] Similarly, various strains of microfungi inhabiting various regions in Victoria, continental Antarctica, have been screened for the production of enzymes, such as amylase, cellulases, proteases, lipases, pectinase, polygalacturonase, phosphatase, glucose oxidase, chitinase, urease, DNase and RNase.[111] Similarly, keratinases have been found to be produced by different fungal strains, such as *Geomyces pannorum* from Antarctica.[54] Also, species of microfungi such as *Penicillium*, *Cadophora*, *Cladosporium* and *Geomyces* have been found to produce extracellular endo-1,4-β-glucanase at 15 °C and 4 °C.[132] On the other hand, Damare and coworkers (2006)[133] reported *Aspergillus ustus* as the highest producer of cold-tolerant protease from deep-sea mycoflora. Similarly, polygalacturonase enzymes from deep-sea yeast and alkaline proteases from fungi have also been characterised.[133] Various research groups, e.g., Zhu and coworkers (2009) [134] and Turchett and coworkers (2008),[135] have reported cold-adapted enzymes from microfungi.

6.7 CONCLUSION

The extensive efforts that have been carried out worldwide to understand the diversity and distribution of extremophilic microbes have revealed that the knowledge gained so far is just the tip of the iceberg. The application of metagenomic and metaproteomic approaches has already provided valuable insights into biological functionality in marine Antarctic ecosystems and can be similarly applied to terrestrial ecosystems to understand these systems. To date, only a limited number of microfungi have been successfully isolated and cultured from cold environments. Improvement is needed in the culturing methods and growth culture media to isolate a greater diversity of microfungi having unique physiologies to adapt themselves in such extreme environments. It would be advantageous to have important insights into the metabolic potential of these cold-inhibiting microbes and their accurate quantification for further applications in diverse sectors.

6.8 FUTURE PERSPECTIVES

A large number of valuable and more novel extremophilic microorganisms are likely to be discovered in the near future, and cold-adapted microbes represent one such category of extremophilic microbes. Focused research on extensive and diverse culture methods to isolate cold-adapted microfungi will be required to establish an efficient culture collection, since only sporadic efforts have been made to isolate these microfungi from cold habitats, particularly that of the Indian subcontinent. Cold-inhibiting microbes have great ecological and physiological significance, leading to the discovery of various novel and unexplored bioactive metabolites, which will unravel their mechanism of action and help to benefit mankind in diverse ways, i.e., in the pharmaceutical, biotechnological and agricultural fields. Extensive funding, concerted, collaborative endeavours, and interdisciplinary research are needed among academia (biologists and chemists), industry and government.

ACKNOWLEDGEMENT

No acknowledgement.

REFERENCES

1. Margesin, Rosa, and Vanya Miteva. "Diversity and ecology of psychrophilic microorganisms." *Research in Microbiology* 162 no. 3 (2011): 346–361.
2. Cavicchioli, Ricardo, Khawar S. Siddiqui, David Andrews et al. Sowers. "Low-temperature extremophiles and their applications." *Current Opinion in Biotechnology* 13 no. 3 (2002): 253–261.
3. Raspor, Peter, and Jure Zupan. "Yeasts in extreme environments." in *Biodiversity and Ecophysiology of Yeasts*, 371–417, Springer, Berlin, Heidelberg, 2006.
4. Widden, Paul. "Fungal communities in soils along an elevation gradient in northern England." *Mycologia* 79 no. 2 (1987): 298–309.
5. S. Ram Upadhyay and Rai Bharat. "Coprinus lagopus as potent saprophytic colonizer of pigeonpea in soil." *Science and Culture* (1979).
6. Julian Howell Miller, Joel Edwin Giddens, and A. A. Foster. "A survey of the fungi of forest and cultivated soils of Georgia." *Mycologia* 49 no. 6 (1957): 779–808.
7. T. K. Ramachandra Reddy. "Role of plant cover in distribution of fungi in Nilgiri forest soils." in *Proceedings/Indian Academy of Sciences*, Springer, India 56 no. 4 (1962): 185–194.
8. Skarma Nonzom, and Sumbali Geeta. "Dominance of monoverticillate penicilli in the cold arid soils of Drass valley, India." *Studies in Fungi* 6 no. 1 (2021): 378–390.
9. J. Cynan Ellis-Evans. "Microbial diversity and function in Antarctic freshwater ecosystems." *Biodiversity & Conservation* 5 no. 11 (1996): 1395–1431.
10. Brett E. Arenz, Benjamin W. Held, Joel A. Jurgens et al. "Fungal diversity in soils and historic wood from the Ross Sea Region of Antarctica." *Soil Biology and Biochemistry* 38 no. 10 (2006): 3057–3064.

11. Kochkina A. Galina, Ivanushkina E. Nataliya, Lupachev V. Alexey et al. "Diversity of mycelial fungi in natural and human-affected Antarctic soils." *Polar Biology* 42 no. 1 (2019): 47–64.
12. Loque, Carolina P., Adriana O. Medeiros, Franciane M. Pellizzari et al. "Fungal community associated with marine macroalgae from Antarctica." *Polar Biology* 33 no. 5 (2010): 641–648.
13. Christensen, Martha. "Species diversity and dominance in fungal community. Carroll GW Wicklow DT The fungal community, its organization and role in the ecosystem." (1981).
14. El-Said, H. M. Ahmed. "Studies on soil mycoflora of Bahreen." *Microbiological Research* 149 no. 3 (1994): 263–269.
15. Skarma, Nonzom and Sumbali Geeta. "Incidence and diversity spectrum of mitosporic fungi from the cold arid base soil of Moonland landscape (Ladakh), India." *International Journal of Pharma and Bio Sciences* 6 no. 1 (2015).
16. Buzzini, Pietro, Eva Branda, Marta Goretti et al. "Psychrophilic yeasts from worldwide glacial habitats: Diversity, adaptation strategies and biotechnological potential." *FEMS Microbiology Ecology* 82 no. 2 (2012): 217–241.
17. K. Sunil Deshmukh "Isolation of dermatophytes and other keratinophilic fungi from soil of Mussoorie (India)." *Mycoses* 28 no. 2 (1985): 98–101.
18. K. Sunil Deshmukh, P. C. Jain, and S. C. Agrawal. "*Chrysosporium merdarium* (Link) Carmichael: A new record from India." *Acta Botanica Indica* (1985).
19. Christopher W. Schadt, Renée B. Mullen, and Steven K. Schmidt. "Isolation and phylogenetic identification of a dark-septate fungus associated with the alpine plant *Ranunculus adoneus*." *New Phytologist* 150 no. 3 (2001): 747–755.
20. Sunil K. Deshmukh "Incidence of dermatophytes and other keratinophilic fungi in the glacier bank soils of the Kashmir valley, India." *Mycologist* 16, no. 4 (2002): 165–167.
21. Sunil K. Deshmukh, Shilpa A. Verekar, and Archana Shrivastav. "The occurrence of keratinophilic fungi in selected soils of Ladakh (India)." *Natural Science* 2 no. 11 (2006): 1247.
22. Oliver Mühlmann and Ursula Peintner. "Mycobionts of Salix herbacea on a glacier forefront in the Austrian Alps." *Mycorrhiza* 18 no. 4 (2008): 171–180.
23. K. R. Freeman, A. P. Martin, D. Karki, R. C. Lynch, M. S. Mitter, A. F. Meyer, J. E. Longcore, D. R. Simmons, and S. K. Schmidt. "Evidence that chytrids dominate fungal communities in high-elevation soils." *Proceedings of the National Academy of Sciences* 106 no. 43 (2009): 18315–18320.
24. Shona M. Duncan, Roberta L. Farrell, Neville Jordan, Joel A. Jurgens, and Robert A. Blanchette. "Monitoring and identification of airborne fungi at historic locations on Ross Island, Antarctica." *Polar Science* 4 no. 2 (2010): 275–283.
25. K. Sunil Deshmukh, Shilpa A. Verekar, and Archana Shrivastav. "The occurrence of keratinophilic fungi in selected soils of Ladakh (India)." *Natural science* 2 no. 11 (2010): 1247.
26. Fujiyoshi Masaaki, Shinpei Yoshitake, Kenichi Watanabe et al. "Successional changes in ectomycorrhizal fungi associated with the polar willow *Salix polaris* in a deglaciated area in the High Arctic, Svalbard." *Polar Biology* 34 no. 5 (2011): 667–673.
27. Li Hai-Yan, Mi Shen, Zuo-Ping Zhou et al. "Diversity and cold adaptation of endophytic fungi from five dominant plant species collected from the Baima Snow Mountain, Southwest China." *Fungal Diversity* 54 no. 1 (2012): 79–86.
28. Lamabam Sophiya Devi, Polashree Khaund, Fenella MW Nongkhlaw et al. "Diversity of culturable soil micro-fungi along altitudinal gradients of Eastern Himalayas." *Mycobiology* 40 no. 3 (2012): 151–158.

29. Priyanka Sati, Kusum Dhakar, and Anita Pandey. "Microbial diversity in soil under potato cultivation from cold desert Himalaya, India." *International Scholarly Research Notices* 2013 (2013).
30. Hassan, Noor. "Isolation and characterization of psychrophilic fungi from Batura Passu and Siachen Glaciers, Pakistan." PhD diss., Dissertation. Department of Microbiology, Quaid-i-Azam University, Islamabad, 2015.
31. Frey Beat, Thomas Rime, Marcia Phillips et al. "Microbial diversity in European alpine permafrost and active layers." *FEMS microbiology ecology* 92, no. 3 (2016): fiw018.
32. Singh M. Shiv, Masaharu Tsuji, Puja Gawas-Sakhalker et al. "Bird feather fungi from Svalbard Arctic." *Polar Biology* 39 no. 3 (2016): 523–532.
33. Muggia Lucia, Antonia Fleischhacker, Theodora Kopun et al. "Extremotolerant fungi from alpine rock lichens and their phylogenetic relationships." *Fungal Diversity* 76 no. 1 (2016): 119–142.
34. Wang Manman, Jianqing Tian, Meichun Xiang et al. "Living strategy of cold-adapted fungi with the reference to several representative species." *Mycology* 8 no. 3 (2017): 178–188.
35. Noor Hassan, Muhammad Rafiq, Muhammad Hayat et al. "Potential of psychrotrophic fungi isolated from Siachen glacier, Pakistan, to produce antimicrobial metabolites." *Appl Ecol Environ Res* 15 no. 3 (2017): 1157–1171.
36. Adamczyk Magdalene, Frank Hagedorn, Sonja Wipf et al. "The soil microbiome of Gloria Mountain summits in the Swiss Alps." *Frontiers in Microbiology* 10 (2019): 1080.
37. Xie Fei, Anzhou Ma, Hanchang Zhou et al. "Revealing fungal communities in alpine wetlands through species diversity, functional diversity and ecological network diversity." *Microorganisms* 8 no. 5 (2020): 632.
38. Angela Yaneth Landínez-Torres, Jessika Lucia Becerra Abril et al. "Soil microfungi of the colombian natural regions." *International Journal of Environmental Research and Public Health* 17 no. 22 (2020): 8311.
39. Jain Rahul, Bhardwaj Priyanka, Pandey Shiv Shanker et al. "*Arnebia euchroma*, a plant species of cold desert in the Himalayas, harbors beneficial cultivable endophytes in roots and leaves." *Frontiers in Microbiology* 12 (2021).
40. Kluge Mariana, Maxime Wauthy, Karina Engelbrecht Clemmensen et al. "Declining fungal diversity in Arctic freshwaters along a permafrost thaw gradient." *Global Change Biology* 27 no. 22 (2021): 5889–5906.
41. Ogórek Rafał, Jakub Suchodolski, Agata Piecuch et al. "Keratinophilic and Keratinolytic fungi in cave ecosystems: A culture-based study of Brestovská cave and Demänovská Ľadová and Slobody caves (Slovakia)." *Applied Sciences* 12 no. 3 (2022): 1455.
42. Tamara Martin-Pozas, Alena Nováková, Valme Jurado et al. "Diversity of microfungi in a high radon cave ecosystem." *Frontiers in Microbiology* 13 (2022).
43. Chen Jian, Zuomin Shi, Shun Liu et al. "Altitudinal variation influences soil fungal community composition and diversity in Alpine–Gorge Region on the Eastern Qinghai–Tibetan Plateau." *Journal of Fungi* 8 no. 8 (2022): 807.
44. V. A. Iliushin, IYu Kirtsideli, and N. A. Sazanova. "Diversity of microfungi of coal mine spoil tips in the Magadan Region, Russia." *Current Research in Environmental & Applied Mycology (Journal of Fungal Biology)* 12 no. 1 (2022): 136–146.
45. Tosi Solveig, Begoña Casado, Renato Gerdol et al. "Fungi isolated from Antarctic mosses." *Polar Biology* 25, no. 4 (2002): 262–268.
46. I. Wladyslav Golubev. "New species of basidiomycetous yeasts, *Rhodotorula creatinovora* and *R. yakutica*, isolated from permafrost soils of Eastern-Siberian Arctic." *Mykologiya i Phytopathologiya* 32 (1998): 8–13.

47. Ma Li-Jun, Scott O. Rogers, Catharine M. Catranis et al. "Detection and characterization of ancient fungi entrapped in glacial ice." *Mycologia* 92 no. 2 (2000): 286–295.
48. A. Paul Broady and Weinstein Richard N. "Algae, lichens and fungi in la Gorce Mountains, Antarctica." *Antarctic Science* 10 no. 4 (1998): 376–385.
49. Blanchette A. Robert, Benjamin W. Held, Jurgens A. Joel et al. "Wood-destroying soft rot fungi in the historic expedition huts of Antarctica." *Applied and Environmental Microbiology* 70 no. 3 (2004): 1328–1335.
50. Vishniac S. Helen. "The microbiology of Antarctic soils." In *Antarctic Microbiology*, 297–342, Wiley-Liss Inc., New York: Friedmann, 1993.
51. Cameron E. Roy, Richard C. Honour, and Frank A. Morelli. "Antarctic microbiology–preparation for Mars life detection, quarantine, and back contamination." in *Extreme Environments*, Academic Press, 1976, 57–82.
52. Bergero Roberta, Girlanda Mariangela, Varese C. Giovanna et al. "Psychrooligotrophic fungi from arctic soils of Franz Joseph Land." *Polar biology* 21 no. 6 (1999): 361–368.
53. Rosa Luiz Henrique, Vieira D.L.A. Mariana, Santiago F. Lara et al. "Endophytic fungi community associated with the dicotyledonous plant *Colobanthus quitensis* (Kunth) Bartl. (Caryophyllaceae) in Antarctica." *FEMS Microbiology Ecology* 73 no. 1 (2010): 178–189.
54. Marshall A. William. "Aerial transport of keratinaceous substrate and distribution of the fungus *Geomyces pannorum* in Antarctic soils." *Microbial Ecology* 36 no. 2 (1998): 212–219.
55. S. Helen Vishniac "Biodiversity of yeasts and filamentous microfungi in terrestrial Antarctic ecosystems." *Biodiversity & Conservation* 5 no. 11 (1996): 1365–1378.
56. Mercantini Rinaldo, Marsella Renato, and M. C. Cervellati. "Keratinophilic fungi isolated from Antarctic soil." *Mycopathologia* 106 no. 1 (1989): 47–52.
57. Mercantini Rinaldo, Marsella Renato, Moretto Domenico et al. "Keratinophilic fungi in the Antarctic environment." *Mycopathologia* 122 no. 3 (1993): 169–175.
58. J. R. Sharma "Mycological studies at Antarctica." (2000).
59. Stchigel M. Alberto, C. A. N. O Josep, Walter Mac Cormack et al. "*Antarctomyces psychrotrophicus* gen. et sp. nov., a new ascomycete from Antarctica." *Mycological Research* 105 no. 3 (2001): 377–382.
60. Tosi Solveig, Begoña Casado, Renato Gerdol et al. "Fungi isolated from Antarctic mosses." *Polar Biology* 25 no. 4 (2002): 262–268.
61. Scholler Markus, Martin Schnittler and Meike Piepenbring. "Species of Anthracoidea (Ustilaginales, Basidiomycota) on Cyperaceae in Arctic Europe." *Nova Hedwigia* 76 no. 3–4 (2003): 415–428.
62. Sonjak Silva, Jens C. Frisvad, and Nina Gunde-Cimerman. "Penicillium mycobiota in Arctic subglacial ice." *Microbial Ecology* 52 no. 2 (2006): 207–216.
63. Rao Subramanya, Yuki Chan, Donnabella C. Lacap et al. "Low-diversity fungal assemblage in an Antarctic Dry Valleys soil." *Polar Biology* 35 no. 4 (2012): 567–574.
64. Gonçalves N. Vívian, Aline BM Vaz, Carlos A. Rosa et al. "Diversity and distribution of fungal communities in lakes of Antarctica." *FEMS Microbiology Ecology* 82 no. 2 (2012): 459–471.
65. Convey Peter, Hafizah Siti Hafizah, Alias Siti Aisyah et al. "Studies on diversity of soil microfungi in the Hornsund area, Spitsbergen." *Polish Polar Research* 1 (2013): 39–54.

66. Zhang Tao, Yu-Qin Zhang, Hong-Yu Liu et al. "Diversity and cold adaptation of culturable endophytic fungi from bryophytes in the Fildes Region, King George Island, maritime Antarctica." *FEMS Microbiology Letters* 341 no. 1 (2013): 52–61.
67. Santiago F. Iara, Marco Aurélio Soares, Carlos A. Rosa et al. "Lichensphere: A protected natural microhabitat of the non-lichenised fungal communities living in extreme environments of Antarctica." *Extremophiles* 19 no. 6 (2015): 1087–1097.
68. Zhang Tao and Yi-Feng Yao. "Endophytic fungal communities associated with vascular plants in the high arctic zone are highly diverse and host-plant specific." *PLoS One* 10 no. 6 (2015): e0130051.
69. de Menezes, Graciéle CA, Valéria M. Godinho et al. "Antarctomyces pellizariae sp. nov., a new, endemic, blue, snow resident psychrophilic ascomycete fungus from Antarctica." *Extremophiles* 21 no. 2 (2017): 259–269.
70. Robicheau M. Brent, Sarah J. Adams, Jennifer F. Provencher et al. "Diversity and keratin degrading ability of fungi isolated from Canadian Arctic marine bird feathers." *Arctic* 72 no. 4 (2019): 347–359.
71. Perini Laura, Cene Gostinčar, and Nina Gunde-Cimerman. "Fungal and bacterial diversity of Svalbard subglacial ice." *Scientific Reports* 9 no. 1 (2019): 1–15.
72. Rosa Luiz Henrique, Thamar Holanda da Silva, Mayara Baptistucci Ogaki et al. "DNA metabarcoding uncovers fungal diversity in soils of protected and non-protected areas on Deception Island, Antarctica." *Scientific Reports* 10, no. 1 (2020): 1–9.
73. Newsham K. Kevin, Marie L. Davey, David W. Hopkins et al. "Regional diversity of maritime Antarctic soil fungi and predicted responses of guilds and growth forms to climate change." *Frontiers in Microbiology* 11 (2021): 615659.
74. de Souza Láuren, Mayara B. Ogaki, Paulo EAS Câmara et al. Assessment of fungal diversity present in lakes of Maritime Antarctica using DNA metabarcoding: A temporal microcosm experiment." *Extremophiles* 25 no. 1 (2021): 77–84.
75. Varrella Stefano, Giulio Barone, Michael Tangherlini et al. "Diversity, ecological role and biotechnological potential of antarctic marine fungi." *Journal of Fungi* 7 no. 5 (2021): 391.
76. Smith E. Sally and David J. Read. *Mycorrhizal Symbiosis*, 1–769. New York: Academic Press, 2010.
77. Deslippe Julie, Hartmann Martin, Mohn W. William et al. "Long-term experimental manipulation of climate alters the ectomycorrhizal community of Betula nana in Arctic tundra." *Global Change Biology* 17 no. 4 (2011): 1625–1636.
78. Dhakar Kusum and Pandey Anita. "Microbial ecology from the himalayan cryosphere perspective." *Microorganisms* 8 no. 2 (2020): 257.
79. Inman Mason. "The story 'Settling the Science on Himalayan Glaciers'." *Nature* (2010): 28–30.
80. Sati C. Sushil and Belwal Mukta. "Aquatic hyphomycetes as endophytes of riparian plant roots." *Mycologia* 97 no. 1 (2005): 45–49.
81. Lone A. Museer, Hamid Burhan, Singh Poonam et al. "Isolation and characterization of fungal species from the rhizosphere of Juglans regia L. of Kashmir Valley." *Int J Inst Pharm Life Sci* 1 no. 2 (2011): 18–27.
82. Sagar Anand and Raghwa Smriti. "Studies on the mycoflora of cold desert area of Himachal Pradesh." *Indian Phytopathology* (2011): 35–41.
83. Shivaji Sisinthy, Bhaskar Bhadra, Ravella Sreenivas Rao et al. "*Rhodotorula himalayensis* sp. nov., a novel psychrophilic yeast isolated from Roopkund Lake of the Himalayan Mountain ranges, India." *Extremophiles* 12 no. 3 (2008): 375–381.

84. Nadeem, Shaukat. Diversity of culturable psychrophilic fungi from Tirich Mir glacier, Chitral. Dissertation, Department of Microbiology, Quaid-i-Azam University, Islamabad, 2014.
85. Wang Mingjun, Jiang Xiangao, Wu Wenping et al. "Psychrophilic fungi from the world's roof." *Persoonia-Molecular Phylogeny and Evolution of Fungi* 34 no. 1 (2015): 100–112.
86. Dhakar Kusum, Sharma Avinash and Pandey Anita. "Cold, pH and salt tolerant Penicillium spp. inhabit the high-altitude soils in Himalaya, India." *World Journal of Microbiology and Biotechnology* 30 no. 4 (2014): 1315–1324.
87. Sandeep Kotwal and Geeta Sumbali. "Incidence of myco-keratinophiles in cold arid soil at high altitude Khardung village of Ladakh, India." *J Mycol Plant Pathol.* 41 (2011): 72–76.
88. Anupama D. Pulikkal, Kumar D. Praveen, Singh K. Ram. et al. "A psychrophilic and halotolerant strain of Thelebolus microsporus from Pangong Lake, Himalaya." *Mycosp* 2 no. 5 (2011): 601–609.
89. Nonzom, Skarma and Sumbali Geeta. "New record of Chaetasbolisia erysiphoides from cold arid soils of Zanskar (Kargil), India." *Current Science* 114 no. 1 (2018): 25–27.
90. Gupta Puja, Naseer Sangwan, Rup Lal et al. "Bacterial diversity of Drass, cold desert in Western Himalaya, and its comparison with Antarctic and Arctic." *Archives of Microbiology* 197 no. 6 (2015): 851–860.
91. Gupta Puja and Vakhlu Jyoti. "Culturable bacterial diversity and hydrolytic enzymes from drass, a cold desert in India." *African Journal of Microbiology Research* 9 no. 31 (2015): 1866–1876.
92. Ruisi Serena, Donatella Barreca, Laura Selbmann et al. "Fungi in Antarctica." *Reviews in Environmental Science and Bio/Technology* 6 no. 1 (2007): 127–141.
93. J. Nicholas Russell "Cold adaptation of microorganisms." *Philosophical Transactions of the Royal Society of London. B, Biological Sciences* 326 no. 1237 (1990): 595–611.
94. Crowe H. John, Folkert A. Hoekstra and Lois M. Crowe. "Anhydrobiosis." *Annual Review of Physiology* 54 no. 1 (1992): 579–599.
95. A. David Ratkowsky and Genevieve M. Gates. "An inventory of macrofungi observed in Tasmanian forests over a six-year period." *Tasforests-Hobart* 16 (2005): 153.
96. Krishnan Abiramy, Peter Convey, Marcelo Gonzalez et al. "Effects of temperature on extracellular hydrolase enzymes from soil microfungi." *Polar Biology* 41 no. 3 (2018): 537–551.
97. Pacelli Claudia, Laura Selbmann, Ralf Moeller et al. "Cryptoendolithic Antarctic black fungus *Cryomyces antarcticus* irradiated with accelerated helium ions: Survival and metabolic activity, DNA and ultrastructural damage." *Frontiers in Microbiology* 8 (2017): 2002.
98. Coleine Claudia, Laura Zucconi, Silvano Onofri et al. "Sun exposure shapes functional grouping of fungi in cryptoendolithic Antarctic communities." *Life* 8 no. 2 (2018): 19.
99. Crowe H. John, Lois M. Crowe, John F. Carpenter et al. "Stabilization of dry phospholipid bilayers and proteins by sugars." *Biochemical Journal* 242 no. 1 (1987): 1.
100. Jennings H. David "Polyol metabolism in fungi." *Advances in Microbial Physiology* 25 (1985): 149–193.
101. D'Amore Tony, Rena Crumplen and Graham G. Stewart. "The involvement of trehalose in yeast stress tolerance." *Journal of Industrial Microbiology* 7 no. 3 (1991): 191–195.

102. Niederer Markus, Pankow Wulf and Wiemken Andres. "Seasonal changes of soluble carbohydrates in mycorrhizas of Norway spruce and changes induced by exposure to frost and desiccation." *European Journal of Forest Pathology* 22 no. 5 (1992): 291–299.
103. Weinstein N. Richard, Pedro O. Montiel and Johnstone Keith. "Influence of growth temperature on lipid and soluble carbohydrate synthesis by fungi isolated from fell-field soil in the maritime Antarctic." *Mycologia* 92 no. 2 (2000): 222–229.
104. Tibbett K. Mark, Grantham F. E. Sanders and Cairney W. G. John. "Induction of cold active acid phosphomonoesterase activity at low temperature in psychrotrophic ectomycorrhizal Hebeloma spp." *Mycological Research* 102 no. 12 (1998): 1533–1539.
105. Hoshino Tamotsu. "Ecophysiology of snow mold fungi." *Current Topics in Plant Biology* 6 (2005): 27–35.
106. Singh, M. Shiv., Gawas Puja, and Bhat J. Darbhe. "Psychrophilic fungi from Schirmacher Oasis, East Antarctica." *Current Science* (2006): 1388–1392.
107. Gerday Charles, Mohamed Aittaleb, Jean Louis Arpigny et al. "Psychrophilic enzymes: A thermodynamic challenge." *Biochimica et Biophysica Acta (BBA)-Protein Structure and Molecular Enzymology* 1342, no. 2 (1997): 119–131.
108. Ray K. Malay, Uma K. Devi, Kumar G. Seshu et al. "Extracellular protease from the Antarctic yeast *Candida humicola*." *Applied and Environmental Microbiology* 58 no. 6 (1992): 1918–1923.
109. Duncan M. Shona, Ryuji Minasaki, Roberta L. Farrell et al. "Screening fungi isolated from historic Discovery Hut on Ross Island, Antarctica for cellulose degradation." *Antarctic Science* 20 no. 5 (2008): 463–470.
110. Gesheva, Victoria. "Distribution of psychrophilic microorganisms in soils of Terra Nova Bay and Edmonson Point, Victoria Land and their biosynthetic capabilities." *Polar Biology* 32 no. 9 (2009): 1287–1291.
111. Broady Paul, David Given, Laurence Greenfield et al. "The biota and environment of fumaroles on Mt Melbourne, Northern Victoria Land." *Polar Biology* 7 no. 2 (1987): 97–113.
112. Ding Zhuang, Li Liyuan, Che Qian et al. "Richness and bioactivity of culturable soil fungi from the Fildes Peninsula, Antarctica." *Extremophiles* 20 no. 4 (2016): 425–435.
113. Li Yan, Bingda Sun, Shuchun Liu et al. "Bioactive asterric acid derivatives from the Antarctic ascomycete fungus *Geomyces* sp." *Journal of Natural Products* 71 no. 9 (2008): 1643–1646.
114. Li Liyuan, Dehai Li, Yepeng Luan et al. "Cytotoxic metabolites from the antarctic psychrophilic fungus *Oidiodendron truncatum*." *Journal of Natural Products* 75 no. 5 (2012): 920–927.
115. Gonçalves N. Vivian, Carvalho R. Camila, Johann Susana et al. "Antibacterial, antifungal and antiprotozoal activities of fungal communities present in different substrates from Antarctica." *Polar Biology* 38 no. 8 (2015): 1143–1152.
116. Godinho M. Valéria, Vívian N. Gonçalves, Iara F. Santiago et al. "Diversity and bioprospection of fungal community present in oligotrophic soil of continental Antarctica." *Extremophiles* 19 no. 3 (2015): 585–596.
117. de Menezes Graciéle Cunha Alves, Bárbara Alves Porto, Soraya Sander Amorim et al. "Fungi in glacial ice of Antarctica: Diversity, distribution and bioprospecting of bioactive compounds." *Extremophiles* 24, no. 3 (2020): 367–376.
118. G. Anatoly Kozlovsky, Kochkina A. Galina, Zhelifonova P. Valentina et al. "Secondary metabolites of the genus Penicillium from undisturbed and anthropogenically altered Antarctic habitats." *Folia Microbiologica* 65 no. 1 (2020): 95–102.

119. Anne-Monique Gounot. "Bacterial life at low temperature: Physiological aspects and biotechnological implications." *Journal of Applied Bacteriology* 71 no. 5 (1991): 386–397.
120. Huston L. Adrienne, Barbara B. Krieger-Brockett and Jody W. Deming. "Remarkably low temperature optima for extracellular enzyme activity from Arctic bacteria and sea ice." *Environmental Microbiology* 2 no. 4 (2000): 383–388.
121. Georlette Daphné, Benjamin Damien and Vinciane Blaise. "Structural and functional adaptations to extreme temperatures in psychrophilic, mesophilic, and thermophilic DNA ligases." *Journal of Biological Chemistry* 278 no. 39 (2003): 37015–37023.
122. Feller Urs, Steven J. Crafts-Brandner and Michael E. Salvucci. "Moderately high temperatures inhibit ribulose-1, 5-bisphosphate carboxylase/oxygenase (Rubisco) activase-mediated activation of Rubisco." *Plant Physiology* 116 no. 2 (1998): 539–546.
123. Suzuki Takeshi, Toru Nakayama, Tatsuo Kurihara et al. Cold-active lipolytic activity of psychrotrophic Acinetobacter sp. strain no. 6." *Journal of Bioscience and Bioengineering* 92 no. 2 (2001): 144–148.
124. Villeret Vincent, Jozef Van Beeumen, Jean-Pierre Chessa et al. "Preliminary crystal structure determination of the alkaline protease from the Antarctic psychrophile Pseudomonas aeruginosa." *Protein Science* 6 no. 11 (1997): 2462–2464.
125. Tony Collins, Marie-Alice Meuwis, Ingeborg Stals et al. "A novel family 8 xylanase, functional and physicochemical characterization." *Journal of Biological Chemistry* 277, no. 38 (2002): 35133–35139.
126. G. Akila, and T. S. Chandra. "A novel cold-tolerant Clostridium strain PXYL1 isolated from a psychrophilic cattle manure digester that secretes thermolabile xylanase and cellulase." *FEMS Microbiology Letters* 219 no. 1 (2003): 63–67.
127. Mavromatis Konstantinos, Matteo Lorito, Sheridan L. Woo et al. "Mode of action and antifungal properties of two cold-adapted chitinases." *Extremophiles* 7 no. 5 (2003): 385–390.
128. Nakagawa Takahiko, Nagaoka Tsutomu, Taniguchi Seiji et al. "Isolation and characterization of psychrophilic yeasts producing cold-adapted pectinolytic enzymes." *Letters in Applied Microbiology* 38 no. 5 (2004): 383–387.
129. Bradner R. John, Gillings Michael and Nevalainen K. M. H. "Qualitative assessment of hydrolytic activities in Antarctic microfungi grown at different temperatures on solid media." *World Journal of Microbiology and Biotechnology* 15 no. 1 (1999): 131–132.
130. Zucconi Laura, Pagano Sandro, Fenice Massimiliano et al. "Growth temperature preferences of fungal strains from Victoria Land, Antarctica." *Polar Biology* 16 no. 1 (1996): 53–61.
131. Skarma Nonzom and Sumbali Geeta. Studies on the diversity, ecology and enzymatic potential of psychrotolerant microfungi from some unexplored niches of Kargil district (Ladakh). Ph.D. Thesis, Department of Botany, University of Jammu, Jammu, 2018.
132. Duncan M. Shona, Roberta L. Farrell, Joanne M. Thwaites et al. "Endoglucanase-producing fungi isolated from Cape Evans historic expedition hut on Ross Island, Antarctica." *Environmental Microbiology* 8 no. 7 (2006): 1212–1219.
133. Damare Samir, Chandralata Raghukumar, Usha Devi Muraleedharan et al. "Deep-sea fungi as a source of alkaline and cold-tolerant proteases." *Enzyme and Microbial Technology* 39 no. 2 (2006): 172–181.

134. Zhu Hui-Yuan, Yong Tian, Yun-Hua Hou et al. "Purification and characterization of the cold-active alkaline protease from marine cold-adaptive *Penicillium chrysogenum* FS010." *Molecular Biology Reports* 36 no. 8 (2009): 2169–2174.
135. Turchett Benedetta, Pietro Buzzini, Marta Goretti et al. "Psychrophilic yeasts in glacial environments of Alpine glaciers." *FEMS Microbiology Ecology* 63 no. 1 (2008): 73–83.

7 Rhizobiome of the Indian Himalayan ecosystem and its role

Rohini Bhat and Neha Bhat*

CONTENTS

7.1 INTRODUCTION

Cold ecosystems represent the major ecosystem characteristics present in the Earth's biosphere. Both soils and permafrost niches of cold habitats are known to be inhabited by many novel microbes belonging to different domains: archaea (Euryarchaeota), bacteria (Proteobacteria, Actinobacteria and Acidobacter), and fungi (Ascomycota).[1] The Himalayas are one of the critical 'hotspots' of biodiversity of cold-adapted microorganisms. The Indian Himalayan Region (IHR) is a suitable system to study the diversity of the valuable gene pool of psychrophilic microorganisms. Cold-adapted Indian Himalayan microbiomes have been particularly investigated in the last few decades, and many new cold-tolerant microbial species are being added to the list from the IHR region. Soil is known to be the richest repertoire of microbial activity. The rhizosphere is a zone of soil around roots that is characterised by C-rich substances released by roots (such as sugars,

* Corresponding author e-mail: bhatrohini@ymail.com

DOI: 10.1201/9781003354031-7

amino acids and organic acids).[2] Thus, the rhizosphere (rich in organic matter) represents a major 'hot spot' of microbial diversity in soil by providing suitable habitat for the proliferation of microorganisms. These cold-adapted microorganisms conduct a wide range of biosphere services (nutrient cycle, plant health promotion and environmental detoxification) and human economic tasks (production of enzymes). The current chapter discusses the diversity and significance of the cold-adapted rhizosphere microbes of Indian Himalayan regions, which are able, or have potential in the future, to promote the sustainable production of cultivable crops and understand cold adaptation strategies.

7.2 MICROBIOLOGY OF HIMALAYAN RHIZOSPHERIC SOIL

7.2.1 Indian Himalayan Region

The IHR is one of the world's most fragile mountain areas, covering about 5.37 lakh km^2 and accounting for a substantial percentage of a Himalayan biodiversity hotspot.[3] It accounts for 16.2% of India's total land area and is separated into five zones: Trans, North-West, West, Central and East Himalayas. Snow-capped peaks, glaciers and deep forests are the major characteristics of the IHR region. The IHR ecosystem's diverse temperature and physiography result in large variances in microhabitats, which contribute significantly to the richness and representativeness of its floral and faunal assemblage's biodiversity. The Himalayan cold environment is known to house rich gene pools of psychrophilic microbes such as bacteria, archaebacteria and fungi.

7.2.2 Rhizospheric soil

The rhizosphere is a narrow zone of soil around the plant roots characterised by C-rich rhizome deposits (root exudate) that are rich in organic compounds like amino acids, keto acids, simple carbohydrates, glycosides, fatty acids, sterols, vitamins, tannins, alkaloids, phosphatides and secondary metabolites.[4] The general gradients in carbon source, phytochemicals, pH, oxygen and nutrients imposed by the roots result in the enrichment of a percentage of microorganisms (7%) in the bulk soil, and these microbial communities form the rhizosphere microbiome or rhizobiome.[2] The German scientist Hiltner (1904) coined the word *rhizosphere* to describe the interactions between nitrogen-fixing bacteria and the roots of leguminous plants, including how the bacteria consume and fix nitrogen to make it accessible to the plant.[5] Following Hiltner's explanation of the word *rhizosphere*, the concept has been expanded to incorporate all the interactions while frequently appearing to limit its geographical scope.[6] Many archaea (*Haloarcula, Halococcus*), bacteria (*Acinetobacter, Enterobacter, Arthrobacter, Bacillus, Methylobacterium, Flavobacterium, Pseudomonas, Rhizobium, Azospirillum, Burkholderia, Haloferax, Paenibacillus, Serratia*) and fungi (*Aspergillus, Penicillium, Piriformospora*) have been recorded from several agricultural rhizosphere soils.[7]

7.2.3 Bacterial and fungal diversity in Indian Himalayan rhizosphere

7.2.3.1 Bacteria

Due to the small size of the bacteria, they account for a relatively small part of the overall rhizosphere biomass, yet they are the rhizosphere's most abundant inhabitants. Typically, approximately 10^8–10^{12} bacterial cells are present in 1 gram of rhizosphere soil.[8] The standard approach used for evaluating soil bacterial diversity is 16S rRNA gene cloning and sequencing. Gram-negative bacteria predominate in the rhizosphere. Actinobacteria are the dominant inhabitants of the cold soil biotopes due to their metabolic adaptation strategies.[9] The rhizosphere of *Thylacospermum ceaspitosum* was dominated by Actinobacteria, with *Streptomyces*, *Arthrobacter* and *Paenibacillus* as major genera.[10] Soil from the western Himalayas situated at 34.108N, 77.408E was found to be dominated by Proteobacteria (41%), followed by Firmicutes (32%), Actinobacteria (16%) and Bacteroidetes (6%) belonging to genera *Bacillus* (30%) followed by *Pseudomonas* (24%) and *Arthrobacter* (12%).[11] The metagenomic technique was used to study the bacterial diversity of the rhizosphere soil of red kidney beans (*Phaseolus vulgaris*) from the Western Indian Himalaya, and it was discovered that Proteobacteria was the major group, with *Pseudomonas* as the leading genus.[12] *Actinobacteria, Acidobacteria, Bacteroidetes, Chlorobi, Chloroflexi, Cyanobacteria, Firmicutes, Gemmatimonadetes, Nitrospira, Planctomycetes* and *Verrucomicrobia* were also detected. Members of Archaea, cyanobacteria and methanotrophs were observed in the same rhizosphere. Psychrotolerant bacteria associated with wheat from northern hills were found to be dominated by Proteobacteria (65%) followed by Firmicutes (26%), Actinobacteria (8%) and Bacteroidetes (1%) with 45 different species.[13] Rhizosphere bacteria isolated from *Origanum vulgare* in the sub-Himalayan region were dominated by γ-Proteobacteria and Firmicutes.[14] The majority (27.5%) of the isolates belonged to the genus *Pseudomonas*. Rhizospheric soil of established and abandoned tea plants of Indian Himalayan regions was found to be dominated by *Bacillus* (up to 45%) and *Pseudomonas* (up to 85%), respectively.[15] The nitrogen-fixing taxa (Frankia and Cyanobacteria) were more abundant in the rhizospheric soil of *Alnus nepalensis* or Nepali alder of the Darjeeling hills than in the non-rhizospheric or bulk soil.[16]

7.2.3.2 Fungi

Higher plasticity and the capacity of fungi to adapt to different harsh environments make them successful inhabitants of soil.[17] The production of various extracellular enzymes (hydrolases) allows them to break down all types of organic materials and contribute to the maintenance of carbon and nutritional balance. Soil collected from the Himalayas is dominated by Ascomycota and Zygomycota.[18] *Penicillium* predominates among the fungi isolated from higher altitudes of the IHR, followed by *Aspergillus, Paecilomyces, Trichoderma, Cladosporium* and *Phialophora*.[19] These fungal species are characterised by their ability to tolerate a broad range of temperatures, pH and salt concentration. In a study of the Eastern Himalayas carried out by Devi et al.,[18] the isolates from soil belonged to seven orders: Calosphaeriales, Capnodiales, Eurotiales, Hypocreales, Mortierellales, Mucorales and Pleosporales of Ascomycota and Zygomycota. The most common genera belonging to Ascomycota were *Penicillium*, *Talaromyces*, *Chromocleista*, *Paecilomyces*, *Thysanophora*, *Fusarium*, *Trichoderma*, *Hypocrea*, *Aspergillus*, *Emericellopces*, *Nectria* and *Chamaeleomyces*.[19] The most diverse and evenly distributed genera belonging to Zygomycota were *Niesslia*, *Umbelopsis*, *Absidia*, *Mucor*, *Mortierella*, *Penicillium*, *Eurotiales* and *Aspergillus*. The rhizosphere of *Ginkgo biloba* growing in the IHR was dominated by only one genus, *Glomus*.[20] *Glomus*, *Gigaspora*, *Acaulospora*, *Scutellospora*, *Entrophospora* and *Sclerocystis* were reported to be the dominant genera in the rhizosphere of mango orchards growing in the north-western Himalayan region of India.[21] In the trans-Himalayan area, the rhizosphere soils of *Salix alba* Linn. had an abundant population of *Penicillium*, *Aspergillus fumigatus*, *Aspergillus niger* and *Aspergillus* spp.[22] *Acaulospora*, *Glomus*, *Gigaspora*, *Sclerocystis* and *Scutellispora* spores were observed from the rhizosphere soil of central Himalayan rhododendrons, with *Glomus* being the main genus identified from all the rhododendron species.[23] Tea bushes collected from native eco sites of the Kumaun region of Uttarakhand Himalaya were dominated by *Glomus* and *Acaulospora*.[24]

7.3 ROLE OF COLD-ADAPTED RHIZOBIOME

7.3.1 Ecological applications: Plant growth

Microbes from the rhizospheric soil have been shown to promote plant growth in a diversity of ways, including supplying the host plant with critical nutrients found in the soil, such as nitrogen fixation or phosphate solubilisation; modulating plant growth or relieving stress through phytohormone production or degradation; maintaining plant health through competition with phytopathogens or induction of plant resistance; and driving the evolution of multi-disease resistance over time. Because of these roles of the microbiome for the host plant, they are regarded as an important component of the host assemblage and are sometimes referred to as *holobionts*.[25] Many *Bacillus* species have been reported from the soils of the high altitude of the Indian Himalayas. They produce antimicrobial compounds that have antifungal properties against many pathogenic fungi,[23] promote growth and development,[26] and are also involved in nutrient recruitment. Among the Gram-negative bacteria, pseudomonads are known to be the dominant group colonising alpine soils. These are effective disease controllers, since they are well recognised for producing a variety of antimicrobial substances, including 2,4-diacetylphloroglucinol, pyoluteorin, cyclic lipopeptides and hydrolytic enzymes.[27] It has been shown that *Pseudomonas* and *Burkholderia* strains enhance cold tolerance in plants.[28] Inorganic phosphates were found to be solubilised by 49 fluorescent *Pseudomonas* strains, and 91 isolates were shown to inhibit *Rhizoctonia solani* and had antagonistic effect against *Fusarium solani*.[29] In response to cold stress conditions, Pseudomonads present in the rhizosphere of Himalayan plants were shown to accumulate suitable solutes, including sugars, polyols, amino acids, polyamines and raffinose.[30] Pseudomonads from the north-west Himalayan rhizospheric soil enhanced root/shoot biomass and nutrient uptake in wheat.[31] Bacterial species such as *Pseudomonas* sp., *Serratia* sp. and *Rhanella* sp. isolated from the rhizosphere of *Physalis peruviana* L. were shown to possess plant growth-promoting activities.[32] Similarly, microorganisms isolated from the rhizosphere of *Lepidium meyenii* Walp, such as *Pseudomonas* and *Bacillus*, demonstrated features such as indole acetic acid synthesis and phosphate solubilisation.[33] These bacterial strains also help the plant by alleviating stress induced by the environment.[34] Plant growth-promoting activities such as phosphate solubilisation; production of auxin, ammonia and siderophore; and 1-aminocyclopropane-1-carboxylate deaminase activity were reported in *Acinetobacter rhizosphaerae* strain BIHB 723, which was isolated from the rhizosphere of the sea-buckthorn present in the cold deserts of the Himalayas.[35] The strain showed growth-promoting properties under both controlled experiments and field testing. Many bacterial species isolated from the wheat rhizosphere growing in the cold deserts of the Himalayas shared the ability to solubilise phosphate and produce siderophores.[13] Gram-positive and Gram-negative bacteria, actinomycetes and fungi from the rhizospheric soil of potato fields in Mana region (Chamoli district, Uttarakhand) of the Indian Himalayas displayed various plant growth-promoting characteristics, such as indole acetic acid synthesis and phosphate solubilisation.[36] Cold-tolerant *Aspergillus* strains are well recognised for their capacity to solubilise phosphate from aluminium and iron adducts.[37] In the western Himalayas, plant-associated fungal *Trichoderma* species (*T. atroviride*, *T. harzianum*, *T. koningii*, *T. viride* and *T. pseudokoningii*) were shown to possess antagonistic activities against phytopathogens and also possessed plant growth-promoting properties in a plant-based bioassay.[38] Table 7.1 shows some of the important roles played by the dominant microbes present in the IHR rhizospheric soil.

In numerous investigations, rhizospheric soil rich in psychrophilic microorganisms (bacteria and fungi) utilised as inoculum showed a favourable influence on plant development in plant-based bioassays. With some improvements, these microbes can be efficiently used as eco-friendly biofertilisers and biocontrollers in field conditions. An efficient lignocellulolytic consortium of psychrotrophic microbes (*Eupenicillium crustaceum*, *Paceliomyces* sp., *Bacillus atropheus* and *Bacillus* sp.) in combination with some fungal species has been developed to degrade paddy straw and has the potential to degrade agri-residues, enhance soil fertility and decrease pollution.[47]

7.3.2 Biotechnological Applications

Psychrophilic/psychrotolerant microbes produce cold-active enzymes, secondary metabolites and antibiotics and thus can be exploited for industrial applications.[48] These cold-active enzymes play an essential role in the food business (pectinase, galactosidase), textile washing processes and detergent industries (lipases, amylases and cellulases). Many cold-adapted bacteria belonging to *Acinetobacter*, *Aquaspirillium*, *Arthrobacter*, *Bacillus*, *Carnobacterium*,

TABLE 7.1

Prominent microbial groups of the rhizosphere of the Indian Himalayan Region and their role in plant growth

Phylum	Class	Genus	Function
Bacteria (10^8–10^9 g^{-1} soil)			
Proteobacteria	γ-proteobacteria	*Pseudomonas fragi*	P solubilisation, IAA production and HCN production[39]
		Pseudomonas lurida	P solubilisation[40]
		Many strains of *Pseudomonas fluorescens*, *Pseudomonas aeruginosa* and *Pseudomonas putida*	HCN production, siderophore production, P solubilisation and antifungal activity[39,40]
		Pseudomonas vancouverensis	IAA production, P solubilisation, HCN and siderophore production, and antifungal activity[41].
		Rahnella sp.	Inorganic and organic P solubilisation, 1-aminocyclopropane-1-carboxylate deaminase activity, ammonia generation and siderophore production[42]
Actinobacteria	Actinobacteria	*Acinetobacter rhizosphaerae*	Inorganic and organic P solubilisation, auxin and siderophore production, 1-aminocyclopropane-1-carboxylate deaminase activity and ammonia generation[35]
Firmicutes	Bacilli	*Exiguobacterium acetylicum*	P solubilisation, IAA, siderophore and HCN production[40]
		Bacillus amyloliquifaciens	Alkaline protease production[43]
		Bacillus subtilis	Antifungal properties[44]
		Bacillus sp. BPR7	IAA production, siderophore production, production of organic acid, ACC (1-Aminocyclopropane-1-Carboxylate) deaminase cyanogens, lytic enzymes, oxalate oxidase, and solubilisation of P, Zn and K[45]

(*Continued*)

TABLE 7.1 (CONTINUED)

Prominent microbial groups of the rhizosphere of the Indian Himalayan Region and their role in plant growth

Phylum	Class	Genus	Function
Fungi (10^5–10^6 g^{-1} soil)			
Ascomycota		*Penicillium pinophilum*	Prolonged laccase production[46]
		Penicillium spp.	Tricalcium phosphate solubilisation[22]
		Aspergillus fumigatus, *Aspergillus niger*	Tricalcium phosphate solubilisation[22]
Glomeromycota		*Gigaspora*, *Acaulospora*, *Scutellospora*, *Entrophospora*, *Sclerocystis*	Positive effect on shoot extension growth, leaf area, fruit yield, available Cu and Zn content, and leaf N, P, Cu, Zn and Mn contents[21]
		The consortium of species of *Glomus*, *Acaulospora* and *Scutellospora*	Promote the growth of tea plants significantly in nonsterilised acidic soil[24]

HCN, hydrogen cyanide; IAA, indole acetic acid; P, phosphorus

Clostridium, *Cytophaga*, *Flavobacterium*, *Marinomonas*, *Moraxella*, *Moritella*, *Paenibacillus*, *Pantoea*, *Planococcus*, *Pseudoalteromonas*, *Pseudomonas*, *Psychrobacter*, *Shewanella*, *Trichococcus*, *Vibrio* and *Xanthomonas* have been reported to produce cold-active enzymes.[1,48] Antifreeze proteins are found in several psychrophilic bacteria and can be useful for cryo-based surgery and cell/tissue preservation and in the food industry, targeting the preparation of frozen food.[49]

7.4 CONCLUSION AND FUTURE PERSPECTIVE

The Indian Himalayan Region's rhizosphere microbiome appears to be a promising ecological niche for investigating the diversity of cold-adapted microbes, ecosystem functionality and their potential for biotechnological applications. There are already growing numbers of reports on plant growth-promoting properties from the microbes isolated from the cold ecosystem. Thus, there is a greater need for more research on the understanding of the biota in the rhizosphere (metagenomic approach) and their ecology (metabolomics, proteomics and physiological studies). The cold rhizobiome repository can be exploited to select those that have implications for strengthening plant growth and health in diverse stress conditions and thus can largely impact agriculture, horticulture and forestry. Manipulation of the plant root and microbiota interactions can be used for improving plant survival and yield of plant-based products. Cold-active enzymes and secondary metabolites (antibiotics) produced by these microbes open a new potential to many industries.

ACKNOWLEDGEMENT

We gratefully thank the Director, National Centre for Biological Sciences, Bangalore, for providing the necessary facilities and resources for writing this book chapter.

REFERENCES

1. Yadav, Ajar Nath, Priyanka Verma, Vinod Kumar et al. "Extreme cold environments: A suitable niche for selection of novel psychrotrophic microbes for biotechnological applications." *Advances in Biotechnology and Microbiology* 2, no. 2 (2017): 1–4.
2. Pii, Youry, Tanja Mimmo, Nicola Tomasi et al. "Microbial interactions in the rhizosphere: Beneficial influences of plant growth-promoting rhizobacteria on nutrient acquisition process. A review." *Biology and Fertility of Soils* 51, no. 4 (2015): 403–415.
3. Nandy, S. N., P. P. Dhyani, and Prasana K. Samal. *Resource Information Database of the Indian Himalaya*. Environmental Information System on Himalayan Ecology, GP Institute of Himalayan Environment and Development, Almora, 2006.
4. Nowak-Thompson, Brian, Nancy Chaney, Jenny S. Wing et al. "Characterization of the pyoluteorin biosynthetic gene cluster of Pseudomonas fluorescens Pf-5." *Journal of Bacteriology* 181, no. 7 (1999): 2166–2174.

5. Hartmann, Anton, Michael Rothballer, and Michael Schmid. "Lorenz Hiltner, a pioneer in rhizosphere microbial ecology and soil bacteriology research." *Plant and Soil* 312, no. 1 (2008): 7–14.
6. Lettice, Eoin P. "The Rhizosphere: Measuring the zone of interaction." *Annual Plant Reviews online* (2018): 219–236.
7. Kumar, Ashok, and Vijay Singh Meena. "Plant growth promoting rhizobacteria for agricultural sustainability." Springer, Berlin, 10 (2019): 978–981.
8. Kennedy, A. C. "Soil biota in the rhizosphere." In *Principles and Applications of Soil Microbiology*, 242–62. Upper Saddle River, NJ: Prentice Hall, 2005.
9. Makhalanyane, Thulani Peter, Marc Warwick Van Goethem, and Don Arthur Cowan. "Microbial diversity and functional capacity in polar soils." *Current Opinion in Biotechnology* 38 (2016): 159–166.
10. Řeháková, Klára, Alica Chroňáková, Václav Krištůfek et al. "Bacterial community of cushion plant Thylacospermum ceaspitosum on elevational gradient in the Himalayan cold desert." *Frontiers in Microbiology* 6 (2015): 304.
11. Gangwar, Pooja, Syed Imteyaz Alam, Sunita Bansod, and Lokendra Singh. "Bacterial diversity of soil samples from the western Himalayas, India." *Canadian Journal of Microbiology* 55, no. 5 (2009): 564–577.
12. Suyal, Deep Chandra, Amit Yadav, Yogesh Shouche, and Reeta Goel. "Diversified diazotrophs associated with the rhizosphere of Western Indian Himalayan native red kidney beans (*Phaseolus vulgaris* L.)." *3 Biotech* 5, no. 4 (2015): 433–441.
13. Verma, Priyanka, Ajar Nath Yadav, Kazy Sufia Khannam et al. "Assessment of genetic diversity and plant growth promoting attributes of psychrotolerant bacteria allied with wheat (*Triticum aestivum*) from the northern hills zone of India." *Annals of Microbiology* 65, no. 4 (2015): 1885–1899.
14. Bafana, Amit. "Diversity and metabolic potential of culturable root-associated bacteria from Origanum vulgare in sub-Himalayan region." *World Journal of Microbiology and Biotechnology* 29, no. 1 (2013): 63–74.
15. Sood, Anchal, Shivesh Sharma, Vivek Kumar, and Ram Lal Thakur. "Antagonism of dominant bacteria in tea rhizosphere of Indian Himalayan regions." *Journal of Applied Sciences and Environmental Management* 11, no. 4 (2007): 63–66.
16. Sen, Gargi, Indrani Sarkar, Saroja Chhettri et al. "Rhizospheric soil metabarcoding analysis of Alnus nepalensis from Darjeeling hills reveals the abundance of nitrogen-fixing symbiotic microbes." *Journal of Forest Research* 27, no. 2 (2022): 106–112.
17. Frąc, Magdalena, Silja E. Hannula, Marta Bełka, and Małgorzata Jędryczka. "Fungal biodiversity and their role in soil health." *Frontiers in Microbiology* 9 (2018): 707.
18. Devi, Lamabam Sophiya, Polashree Khaund, Fenella MW Nongkhlaw, and S. R. Joshi. "Diversity of culturable soil micro-fungi along altitudinal gradients of Eastern Himalayas." *Mycobiology* 40, no. 3 (2012): 151–158.
19. Pandey, Anita, Kusum Dhakar, Rahul Jain et al. "Cold adapted fungi from Indian Himalaya: Untapped source for bioprospecting." *Proceedings of the National Academy of Sciences, India Section B: Biological Sciences* 89, no. 4 (2019): 1125–1132.
20. Kumar, Ajay, Shipra Singh, and Anita Pandey. "General microflora, arbuscular mycorrhizal colonization and occurrence of endophytes in the rhizosphere of two age groups of *Ginkgo biloba* L. of Indian Central Himalaya." *Indian Journal of Microbiology* 49, no. 2 (2009): 134–141.
21. Sharma, Som Dev, and Pramod Kumar. "Relationship of arbuscular mycorrhizal fungi and Azotobacter with plant growth, fruit yield, soil and leaf nutrient status of

mango orchards in north-western Himalayan region of India." *Journal of Applied Horticulture* 10, no. 2 (2008): 158–163.

22. Chatli, Anshu S., Viraj Beri, and B. S. Sidhu. "Isolation and characterisation of phosphate solubilising microorganisms from the cold desert habitat of Salix alba Linn. in trans Himalayan region of Himachal Pradesh." *Indian Journal of Microbiology* 48, no. 2 (2008): 267–273.
23. Chaurasia, Bhaskar, Anita Pandey, and Lok Man S. Palni. "Distribution, colonization and diversity of arbuscular mycorrhizal fungi associated with central Himalayan rhododendrons." *Forest Ecology and Management* 207, no. 3 (2005): 315–324.
24. Singh, Shipra, Anita Pandey, Bhaskar Chaurasia, and Lok Man S. Palni. "Diversity of arbuscular mycorrhizal fungi associated with the rhizosphere of tea growing in 'natural'and 'cultivated'ecosites." *Biology and Fertility of Soils* 44, no. 3 (2008): 491–500.
25. Vanderkoornhuyse, P., A. Quaiser, M. Duhame et al. "The importance of the microbiome of the plant holobiont." *New Phytologist* 206 (2015): 119.
26. Ghyselinck, Jonas, Siva LS Velivelli, Kim Heylen et al. "Bioprospecting in potato fields in the Central Andean Highlands: Screening of rhizobacteria for plant growth-promoting properties." *Systematic and Applied Microbiology* 36, no. 2 (2013): 116–127.
27. Mishra, Jitendra, and Naveen Kumar Arora. "Secondary metabolites of fluorescent pseudomonads in biocontrol of phytopathogens for sustainable agriculture." *Applied Soil Ecology* 125 (2018): 35–45.
28. Jócsák, Ildikó, György Végvári, and Eszter Vozáry. "Electrical impedance measurement on plants: A review with some insights to other fields." *Theoretical and Experimental Plant Physiology* 31, no. 3 (2019): 359–375.
29. Negi, Yogesh Kumar, Deepti Prabha, Satyendra K. Garg, and J. Kumar. "Genetic diversity among cold-tolerant fluorescent Pseudomonas isolates from Indian Himalayas and their characterization for biocontrol and plant growth-promoting activities." *Journal of Plant Growth Regulation* 30, no. 2 (2011): 128–143.
30. Margesin, Rosa, and Tony Collins. "Microbial ecology of the cryosphere (glacial and permafrost habitats): Current knowledge." *Applied Microbiology and Biotechnology* 103, no. 6 (2019): 2537–2549.
31. McLaughlin, Heather P., Pat G. Casey, Jeni Cotter et al. "Factors affecting survival of Listeria monocytogenes and Listeria innocua in soil samples." *Archives of Microbiology* 193, no. 11 (2011): 775–785.
32. Ogata-Gutiérrez, Katty, Carolina Chumpitaz-Segovia, Jesus Lirio-Paredes et al. "Characterization and potential of plant growth promoting rhizobacteria isolated from native Andean crops." *World Journal of Microbiology and Biotechnology* 33, no. 11 (2017): 1–13.
33. Ortiz-Ojeda, Paola, Katty Ogata-Gutiérrez, and Doris Zúñiga-Dávila. "Evaluation of plant growth promoting activity and heavy metal tolerance of psychrotrophic bacteria associated with maca (Lepidium meyenii Walp.) rhizosphere." *AIMS Microbiology* 3, no. 2 (2017): 279.
34. Rajkumar, M., L. Benedict Bruno, and J. Rajesh Banu. "Alleviation of environmental stress in plants: The role of beneficial *Pseudomonas* spp." *Critical Reviews in Environmental Science and Technology* 47, no. 6 (2017): 372–407.
35. Gulati, Arvind, Pratibha Vyas, Praveen Rahi, and Ramesh Chand Kasana. "Plant growth-promoting and rhizosphere-competent Acinetobacter rhizosphaerae strain BIHB 723 from the cold deserts of the Himalayas." *Current Microbiology* 58, no. 4 (2009): 371–377.

36. Sati, Priyanka, Kusum Dhakar, and Anita Pandey. "Microbial diversity in soil under potato cultivation from cold desert Himalaya, India." *International Scholarly Research Notices* 2013 (2013).
37. Rinu, K., Mukesh Kumar Malviya, Priyanka Sati et al. "Response of cold-tolerant *Aspergillus* spp. to solubilization of Fe and Al phosphate in presence of different nutritional sources." *International Scholarly Research Notices* (2013): 1–10.
38. Ghildiyal, A., and A. Pandey. "Isolation of cold tolerant antifungal strains of *Trichoderma* sp. from glacial sites of Indian Himalayan region." *Research Journal of Microbiology* 3, no. 8 (2008): 559–564.
39. Selvakumar, Govindan, Piyush Joshi, Sehar Nazim et al. "Phosphate solubilization and growth promotion by *Pseudomonas fragi* CS11RH1 (MTCC 8984), a psychrotolerant bacterium isolated from a high altitude Himalayan rhizosphere." *Biologia* 64, no. 2 (2009): 239–245.
40. Selvakumar, Govindan, S. Kundu, Piyush Joshi et al. "Growth promotion of wheat seedlings by Exiguobacterium acetylicum 1P (MTCC 8707) a cold tolerant bacterial strain from the Uttarakhand Himalayas." *Indian Journal of Microbiology* 50, no. 1 (2010): 50–56.
41. Mishra, Pankaj K., Smita Mishra, Govindan Selvakumar et al. "Characterisation of a psychrotolerant plant growth promoting *Pseudomonas* sp. strain PGERs17 (MTCC 9000) isolated from North Western Indian Himalayas." *Annals of Microbiology* 58, no. 4 (2008): 561–568.
42. Vyas, Pratibha, Robin Joshi, K. C. Sharma et al. "Cold-adapted and rhizosphere-competent strain of Rahnella sp. with broad-spectrum plant growth-promotion potential." *Journal of Microbiology and Biotechnology* 20, no. 12 (2010): 1724–1734.
43. Guleria, Shiwani, Abhishek Walia, Anjali Chauhan, and C. K. Shirkot. "Genotypic and phenotypic profile of alkalophile proteolytic bacillus sp. associated with Rhizosphere of Apple Trees in trans Himalayas." *Proceedings of the National Academy of Sciences, India Section B: Biological Sciences* 86, no. 2 (2016): 331–341.
44. Chaurasia, Bhaskar, Anita Pandey, Lok Man S. Palni et al. "Diffusible and volatile compounds produced by an antagonistic *Bacillus subtilis* strain cause structural deformations in pathogenic fungi in vitro." *Microbiological Research* 160, no. 1 (2005): 75–81.
45. Gu, Se Hun, Violaine Nicolas, Aude Lalis et al. "Complete genome sequence and molecular phylogeny of a newfound hantavirus harbored by the Doucet's musk shrew (*Crocidura douceti*) in Guinea." *Infection, Genetics and Evolution* 20 (2013): 118–123.
46. Dhakar, Kusum, Rahul Jain, Sushma Tamta, and Anita Pandey. "Prolonged laccase production by a cold and pH tolerant strain of *Penicillium pinophilum* (MCC 1049) isolated from a low temperature environment." *Enzyme Research* (2014): 1–6.
47. Shukla, Livleen, Archna Suman, Priyanka Verma et al. "Syntrophic microbial system for ex-situ degradation of paddy straw at low temperature under controlled and natural environment." *Journal of Applied Biology and Biotechnology* 4, no. 2 (2016): 0–3.
48. Venkatachalam, Siddarthan, Vasudevan Gowdaman, and Sclai Ramatchandirane Prabagaran. "Culturable and culture-independent bacterial diversity and the prevalence of cold-adapted enzymes from the Himalayan mountain ranges of India and Nepal." *Microbial Ecology* 69, no. 3 (2015): 472–491.
49. Pulicherla, K. K., Mrinimoy Ghosh, P. S. Kumar, and K. R. S. S. Rao. "Psychrozymes-the next generation industrial enzymes." *Journal of Marine Science: Research & Development* 1, no. 2 (2011): 1–7.

8 Soil microbiome for enhanced crop productivity

*Debabrata Das, Koushik Biswas**
and Kumar Abbhishek

CONTENTS

8.1 INTRODUCTION

Soil microorganisms are advantageous in agriculture because they improve plant nutrition and soil health.[1] Efforts are being made to enhance crop production by managing soil microorganisms as a low-input biotechnological tool while retaining the original soil biodiversity.[2] Researchers are focusing on current trends of the root-associated microbiome responsible for economic and ecological sustainability in crop production. In recent years, microbial cultures and their products have been utilised for soil fertility.[3] Based on considerable research into

* Corresponding author e-mail: kb.agriscience@gmail.com; dkb.ag@brainwareuniversity.ac.in

DOI: 10.1201/9781003354031-8

phyto-beneficial microorganisms, several biofertilisers have been created to meet the demand for crop nutrients.[2] According to some greenhouse and field-based research, diverse plant species respond well to microbial inoculation.[4]

Soil comprises huge and varied microbial communities that regulate essential plant growth processes affecting global environmental health. Scientists refer to a collection of microorganisms that play a crucial role in soil respiration and nutrient cycling. Our existing knowledge of the causal relationship between low-input farming and microbial diversity has been challenged by the micro- and macroscale complexity of microbial life in the soil and by erroneous evaluations of its components. Detailed information is necessary to better understand particular soil management or climate-related activities and their repercussions for the categorisation of complex soil microbial diversity.

Nevertheless, recently, as an alternative to comparing genomes extracted from the soil environment, culture-independent methods for analysing the organisation of the soil microbial community have evolved. We now have a better grasp of the soil microbiome due to new sequencing techniques that provide more resolution and coverage. This chapter emphasises the significance of the soil microbiome in the modern world. In addition, specific insights on recent developments in cutting-edge technology to explore hitherto undiscovered soil microbiome components will be discussed. Numerous processes related to plant growth promotion mediated by beneficial microbes have been explained. The most recent paradigms for utilising beneficial microorganisms in diverse stressed agroecosystems are given to develop future insight.

8.2 SOIL MICROBIOME EFFECT ON THE ENVIRONMENT

Soil microbes have direct involvement in greenhouse gas emissions from soil, such as methane (CH_4), carbon dioxide (CO_2), nitric oxide (NO_2) and nitrous oxide (N_2O). Understanding the function of soil bacteria in climate change and how they respond will help determine whether they may be exploited to avoid or trigger a global catastrophe. There may be a function for soil microbes in preventing global warming. Their role in global warming has yet to be defined in a laboratory. The terrestrial forest absorbs a higher quantity of atmospheric carbon dioxide. Soil is one of the most crucial carbon sinks, since it stores twice as much carbon as the atmosphere.[5] Soil bacteria are essential for producing enough food to feed the globe and satisfy global demand. Perhaps this is why scientists are constantly attempting to understand more about the physiology and roles of soil microbes in innovative solutions to the world's food and nutritional security issues. Chemical agents have harmed soil structure and caused undesirable changes in soil microbiology, which negatively impacts climate.[6] Increasingly, probiotic agricultural microorganisms are used to ameliorate climate change and promote plant health to avert these detrimental impacts. In the pursuit of sustainable agriculture, zero-budget farming practices are gaining popularity. Several chemical fertiliser manufacturers have shifted to biofertilisers and other nanotechnology-based chemical fertilisers to preserve soil health and prevent future degradation. IFFCO (Indian

Farmers Fertiliser Cooperative Limited 2019) has just released three nano-based products in India in response to the negative consequences of climate change: Nano Nitrogen, Nano Zinc and Nano Copper. IFFCO reports that such products are environmentally friendly and will reduce the use of chemical fertilisers by up to 50% while increasing crop yield significantly.[7]

A few researchers have also used bacterial species, especially methylotrophic strains, as plant growth stimulants.[8] Methylotrophic bacteria are crucial for mitigating greenhouse gas emissions and decreasing methane, methanol, carbon dioxide and other greenhouse gas production. Organic farming, zero-budget natural farming and other techniques promote using natural fertiliser alternatives rather than synthetic fertilisers to promote soil health over time. Natural approaches help regenerate damaged soil structure and depleted soil microbiota, increasing crop output and providing safe and nutritious food while reducing greenhouse gas emissions. Thus, natural agriculture practices contribute to more consistent soil quality in terms of texture, porosity and microbiota, as well as immobilising toxic substances and preventing them from getting into the food chain. In recent years, liver and renal failure have become more prevalent after consuming meals containing harmful components. In addition, chemical fertilisers contribute to the world's greenhouse gas burden, negatively impacting climate change. Due to these undesired weather fluctuations, which disrupt the biogeochemical cycle, plants and animals suffer. Figure 8.1 depicts alternatives to traditional farming practices for achieving sustainable production and preservation of soil health. Manufacturers of chemical fertilisers have been experimenting with environment- and climate-friendly alternatives.

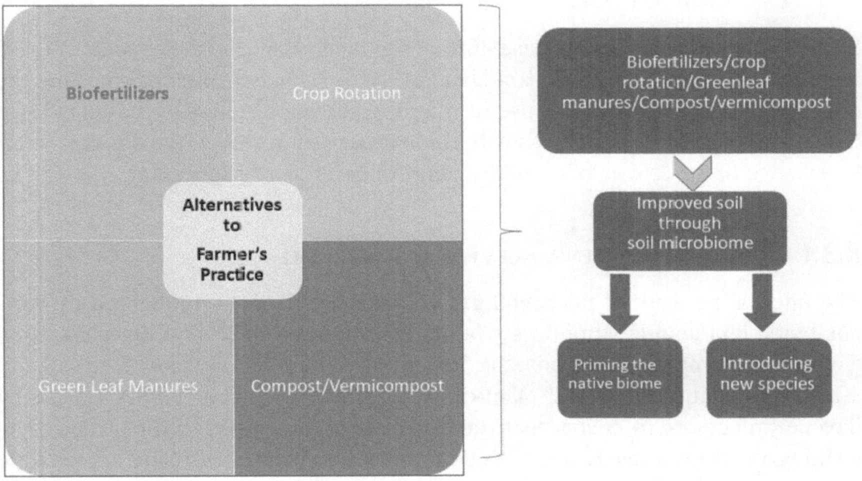

FIGURE 8.1 Alternatives to farmer's practice for achieving sustainable production.

8.2.1 Microbiomes and climate change

Terrestrial plants provide more than 50% of the world's net primary output and generate a substantial amount of biomass.[9] In many terrestrial ecosystems, diverse microbial communities persist based on their geographical and temporal distribution, land use patterns and human disturbances. In the terrestrial environment, plant–microbe interactions involving fungi and bacteria are widespread, but the role of heterotrophic microbial communities in climate change is unknown. As a reliable predictor of the pattern of climate change, a significant store of active carbon in the soil is considered. Since more than 3.5 billion years ago, microorganisms have played an essential role in cycling ozone-depleting compounds in pre- and post-terrestrial contexts. It is extremely likely that climate change will impact microorganisms, just through their interactions with the soil and environment and the presence of ozone-depleting compounds, they may control the atmosphere..[10]

Consequently, global warming affects not just soil's physical and chemical qualities but also the intricate network connections between microbes, plants and animals. These changes are reflected in the responses of microbiomes to their surroundings.[9] Gram-positive and Gram-negative bacteria in the microbial community provide experimental evidence of induced warming.[11] Contradictory findings from short-term laboratory experiments and long-term field investigations on heat adaptation have sparked a discussion on thermal acclimatisation. Soil microbial respiration responds to temperature fluctuations, aided by substrate depletion and decreasing soil biomass, according to studies conducted in the laboratory on a more than 15-years-old midlatitude forest.

8.3 THE BENEFICIAL ROLE OF MICROORGANISMS IN AGRICULTURE

Since agricultural fields comprise 40% of the land surface, the elevation of these regions might result in significant changes to the biogeochemical cycles and the microorganism diversity. The use of microorganisms to mitigate the impacts of climate change on agriculture has become a burning research area, and a better knowledge of the effects of climate change on microbes is necessary.

8.3.1 Plant growth enhancement and microbes

The internal network of microbial growth and development, influenced by various biotic and abiotic stimuli, is wholly dependent on microbial diversity as an element of the natural environment. In contrast, microbial diversity significantly affects the quality of several plant-derived metabolites. The synthesis of many flavone precursors in cocoa beans and tea leaves is regulated, for instance, by forming microflora inside and outside the reaction centre. Utilising a symbiotic or associative fixing process, a subset of rhizobacteria (PGPR) is exceptionally effective at supplying plant roots with a constant supply of critical nutrients. Some of the most crucial features of the function of microbial diversity in the coevolution of agriculture and plant metabolism may be derived for future research. Numerous plant growth-promoting rhizobacteria fix atmospheric and terrestrial nutrient components to plant-usable forms via different mechanisms. Most rhizobacteria have been reported to synthesise auxin series phytohormones and indole acetic acid within their cells, which is thought to play an implicit role in promoting plant growth.

8.3.2 Microbial bioremediation

The role of agricultural microbes is not limited to mobilising the nutrients for plants but has a wide range of applications in bioremediation. Every day, we need clean drinking water, and the demand is increasing exponentially around the globe. Nonetheless, groundwater contamination is increasing with soil pollution, especially from landfills, mining and agrochemical dumping, which makes drinking water scarce for most of the world's population. On the other hand, the activity of soil microflora can be useful in degrading dangerous contaminants and mitigating the effects of inadequate sanitation. Using an efficient bioremediation approach involving *Enterobacter cloacae*, for instance, might significantly decrease selenium pollution in wastewater.[12] Bioremediation using microbial communities is a kind of natural cleansing that entails recycling waste materials into non-toxic or usable forms by employing metabolic events such as eradication, detoxification, degradation and immobilisation. In this context, we will examine bioaugmentation, biostimulation, bioattenuation, biopiling and bioventing, which are gaining acceptance in the scientific community as some of the most promising techniques or approaches. In brief, it is necessary to evaluate possible microbial bioremediation procedures before moving on to soil bacterial bioremediation possibilities.

8.3.3 Bioremediation of heavy metals

Since the buildup and effects of heavy metals on the human body are much more damaging than those of other environmental toxins, pollution caused by heavy metals in soil and water is receiving a great deal of attention. The most toxic heavy metals are strontium (Sr), chromium (Cr), mercury (Hg), lead (Pb), silver (Ag), zinc (Zn), copper (Cu), vanadium (V), tin (Sn), thallium (Tl), cadmium (Cd), arsenic (As), nickel (Ni), cobalt (Co) and caesium (Cs), which provide a threat to ecosystems.[13] Their ionic form has a strong electrostatic affinity for biological components, producing severe damage in the form of molecular destabilisation in essential biomolecules (RNA, DNA and enzymes), which in advanced stages may lead to mutagenesis, cancer or genetic diseases.[14] Multiple studies have found that many microorganisms possess an innate ability to reduce or neutralise the toxicity of most potentially toxic heavy metals by activating processes such as bioaccumulation, bioaugmentation, biosorption, biodegradation and heavy metal resistance (Table 8.1). As a result, future scientific research may look into using

TABLE 8.1
Studies reporting bioremediation using different microbes

Microorganism	Target compounds
Penicillium chrysogenum	Monocyclic hydrocarbons, benzene and phenol compounds[34,35]
Pseudomonas putida	Monocyclic aromatic hydrocarbons[36]
Coprinellus radians	PAHs[37]
Candida viswanathii	Phenanthrene, benzopyrene[38]

various potential genetic engineering techniques that may increase the efficiency of microorganisms in mitigating the adverse effects of heavy metals.

8.3.4 Role of Microbial Flora in the Degradation of PAHs

Microbial populations are essential for elemental recycling, bioremediation of contaminated soils, and waste material turnover. Although volatilisation, adsorption and chemical degradation are all involved in removing poly aromatic hydrocarbons (PAHs) from contaminated soil, the most prevalent degradation mode is microbial degradation, which depends on the populations of microorganisms at the contaminated site and environmental conditions. Hydrocarbon-metabolising bacteria are widespread in polluted environments and play a crucial role in deciding their destiny in all ecosystems.[15] Some microbes, including bacteria, archaea and selective eukaryotes such as fungi and algae, may degrade petroleum's hydrocarbon components. They have a unique ability to decompose petroleum hydrocarbon components by using them as single carbon sources and producing energy in the process. As far as recent research is concerned, 79 bacterial, 9 cyanobacterial, 14 algal and 103 fungal species have been found to break down and alter hydrocarbons.[16]

However, only microorganisms with the necessary enzyme systems can biodegrade petroleum hydrocarbons.[17] Native microbial communities in older polluted soils may be able to effectively metabolise PAHs, which might aid in removing PAHs from freshly polluted soils. Temperature, moisture content, nutrition and *in situ* aeration may influence PAH degradation.

8.4 SOIL MICROBIOME: BENEFACTORS AND THEIR BENEFICIARIES

Soil formation has occurred under various climatic conditions; therefore, individual organisms and their communities may be recognised in various niches in microbial ecology. Soil is a closed system, with nutrient cycling, water management, microbial remediation and biocontrol recognised as major among its activities. According to soil ecosystem services, the microbiome plays a crucial role in various soil activities that influence the long-term productivity of agricultural land.[11] A small number of bacteria that govern most of the soil rhizosphere have a high population density.[18] Rhizospheric research is one of the most exciting fields for investigating microbial communities in soil and plant systems.[19] In the rhizosphere, common microorganisms include fungi, bacteria, protozoa, algae, nematodes and microarthropods. Numerous properties of rhizosphere bacteria allow them to have a direct or indirect positive effect on plant production. Effective soil microbiome management may help us achieve a more outstanding agrarian production over the long run. The diversity of the soil microbiome affects plant growth promotion, phytohormone synthesis and modification, nutrient absorption, and resistance to abiotic and biotic stressors.[20]

8.4.1 Microbial community analysis: Insight into the rhizosphere

The diversity of soil microorganisms is critical for ensuring optimal soil health, since these microbial communities are engaged in several vital processes, including soil formation, N and P elemental cycles, and toxin elimination. Even though there is a great deal of knowledge on the investigation of soil microbial diversity, there are still many challenges to overcome. To overcome these issues, scientists have developed molecular and analytical approaches to enhance the resolving power of soil microbial community composition. Recent advances in next-generation sequencing (NGS) have yielded precise data on the identification and dispersion of microbial species in various environments. Despite hurdles such as microbial population variety and taxonomic uncertainty, high-throughput DNA sequencing revolutionises our understanding of microbiology. New sequencing tools for rhizospheric bacteria are broadening the scope of the species-level study. rDNA intergenic spacer analysis (RISA), DNA cross-hybridisation G+C percent profiling, and community fingerprinting methods such as denaturing gradient gel electrophoresis (DDGE) and terminal restriction fragment length polymorphism (T-RFLP) are the molecular techniques developed to analyse soil microbial communities. Phospholipid fatty acid analysis (PLFA) has been used to monitor microbial populations by tracking key ^{13}C-labelled intermediaries.[21]

8.4.2 Rhizosphere research for sustainable agriculture

The widespread rejection of chemical fertilisers in food production has paved the way for the positive development of biological fertilisers. Additionally, farmers from all around the globe may use this method. In addition, the sustainability of agroecosystems may be improved by conducting extensive research through ecosystem modelling, evaluating all potential outcomes, and collaborating with extension and training networks to disseminate microbial-based agriculture innovations across different socioeconomic classes of the population. Using PGPR in various agroecosystems has enabled a greater understanding of how they work. Several microorganisms, including bacteria and fungi, have been commercialised as a part of such interventions. The microbial interactions in the rhizosphere might be regulated as low-input technologies to promote sustainable agriculture practices. Interactions between members of microbial communities support essential plant growth-promoting activities in an unstable soil environment through actively or indirectly engaging plants in rhizosphere interactions.

8.5 MICROBIAL BIOREMEDIATION: A KEY PLAYER IN CROP PRODUCTIVITY

Biotic and abiotic stress are both considered as major constraints on crop productivity. There is a list of factors that directly or indirectly interfere with the growth and development of a crop species under stress conditions. Of these factors, toxicity, nutritional deficiency, disease and pest susceptibility, hormonal imbalance and physical disorders due to extreme climatic variation are of major concern. On other hand, the impact of such stresses also leads to changes in the carbon, nitrogen and phosphorus cycles and eventually affects the biodiversity of microorganisms. To reduce the impact of such undesirable climatic changes, the application of plant- and animal-associated microorganisms to achieve agricultural production in a sustainable manner has become a rising area of current research. Therefore, a better understanding of the impact of climate change on the microbial community is necessary for mitigating the problem.[22]

In this context, plant–microbe interactions, with special reference to the relationships between soil and the soil microfauna mainly observed in the rhizosphere, are attracting the attention of scientists and researchers for establishing potent rhizoremediation strategies.

8.6 GENETIC EDITING TECHNOLOGY, METAGENOMICS INFLUENCES AND HELP WITH CROP PRODUCTION

As the impact of climate change on agriculture and food security is increasing daily, there is an urgent need for advancement in the development of climate-smart, robust crop plants, which can provide high-value foods with enhanced yield in a sustainable fashion. For the last two decades, plant transformation technology for improved industrial agriculture has played a leading role in developing a few transgenic crop plants. But in most cases, the limitation of transformation and regeneration in many crops makes it challenging for the overall advancement of plant science and technology. High-throughput genome editing, on the other hand, has shown potential to replace the age-old transformation technology, and this could be a revolutionary breakthrough for achieving climate-smart agriculture.[23]

Zinc finger nucleases (ZFNs), a branch of gene editing technology, have undergone extensive forward shaping via programmable nucleases. ZFNs are synthesised proteins used for gene targeting, basically comprising a DNA-cleaving endonuclease bound with zinc finger domains customised to target a precise DNA sequence. Some of the remarkable applications of ZFN-mediated transgene integration are the incorporation of several beneficial traits into maize[24] and targeting safe locations for gene introduction in rice.

Transcription Activator-Like Effector Nucleases (TALENs) are another type of site-directed mutagenesis, which was initially identified in a plant pathogenic bacterium, *Xanthomonas*. Various important genes have been improved by applying the TALEN-based technology in a variety of crop plants, including rice, wheat, tomato, potato and arabidopsis.[25] This technology was first successfully applied in rice by inactivating the OsSWEET14 gene, responsible for bacterial blight susceptibility in the plant.[26] Similarly, a powdery mildew-resistant wheat plant was developed by knocking out three TaMLO homeolog genes.[27]

The applications of the TALEN system in crop improvement programmes are also prominent, with a greater success rate. One gene, caffeic acid O-methyltransferase (COMT), involved in saccharification in sugarcane, has been successfully coedited in 107 copies out of 109, and the mutant species has up to 44% improved saccharification behaviour while producing biofuel without altering its agronomic performance.[28] To improve the nutritional values of soybean with less linoleic acid and greater oleic acid content, knocking out fatty acid desaturase (FAD) genes results in enhancement of shelf life and heat stability of soybean oil.[29]

This technology is considered the most simple, cost-efficient and potent to target multiple genes in a highly precise manner as compared with ZFNs and TALENs . It has been reported that the CRISPR/Cas9 technique has been successfully applied in more than 20 crop species for yield enhancement and increasing resistance to various biotic and abiotic stresses.[30]

Various efforts have been made in developing disease-resistant crop varieties against microbial pathogens by adopting CRISPR/Cas9 technology. A mutated wheat plant resistant to powdery mildew (the causal organism is *Blumeria graminis* sp. *tritici* [Bgt]) was produced by modifying EDR1 homeolog genes applying the CRISPR/Cas9 method.[31] A targeted high degree of resistance against rice blast caused by *Magnaporthe oryzae* was developed by knocking out the OsERF922 gene, employing CRISPR/Cas9 technology, where other key agronomic traits were intact in the modified rice plant.[32] In a recent study, reduced susceptibility to citrus canker in Wanjincheng orange was developed in mutated plants by CRISPR/Cas9-based knocking out of CsWRKY22, a candidate gene for pathogen-induced immunity.[33] Thus, the CRISPRCas9 system has revolutionised the prospects in basic and applied research extensively in the area of crop plant improvement, as this technology has great potential for effortless targeting and huge scope for multiplexed editing of genes in the plant genome.

8.7 CONCLUSIONS

Microorganisms have a wide range of possibilities in terms of their application in nutrient cycling, climate change mitigation, bioremediation, plant protection and carbon sequestration. Microbes in the soil can be used directly through microbial community modulation and niche adaptation or through priming their abilities using genetic engineering tools. The abilities of genetically modified microbes in the rhizosphere to degrade PAHs and immobilise heavy metals are a research landmark. The ecosystem services provided by plant–microbiome interactions seem to have a wide scope for improvement using modern tools in biotechnology. This chapter summarises key processes involved in the rhizospheric dynamics regarding nutrient cycling and bioremediation. We have also listed the key upcoming biotechnological tools that might be useful in engineering the soil microbiota for improved ecosystem services and return from the system. Applying soil microbiota engineering in agricultural production can have a tremendous effect on global food security and climate change mitigation.

ACKNOWLEDGEMENT

Our faculty members of Departmental of Agriculture, Brainware University contributed to writing this chapter.

REFERENCES

1. Lugtenberg, Ben. “Life of microbes in the rhizosphere.” In *Principles of Plant-microbe Interactions*, pp. 7–15. Springer, Cham, 2015.
2. Owen, D., A. P. Williams, G. W. Griffith, and P. J. A. Withers. “Use of commercial bio-inoculants to increase agricultural production through improved phosphrous acquisition.” *Applied Soil Ecology* 86 (2015): 41–54.
3. Vassilev, Nikolay, Eva Martos, Gilberto Mendes et al. “Biochar of animal origin: A sustainable solution to the global problem of high-grade rock phosphate scarcity?” *Journal of the Science of Food and Agriculture* 93, no. 8 (2013): 1799–1804.
4. Calvo, Pamela, Louise Nelson, and Joseph W. Kloepper. “Agricultural uses of plant biostimulants.” *Plant and Soil* 383, no. 1 (2014): 3–41.
5. Sellappan, Ranjith, Aswini Krishnan, and Kalaiselvi Thangavel. “Role of plant microbiome in carbon sequestration for sustainable agriculture.” *Core Microbiome: Improving Crop Quality and Productivity* (2022): 190–205.
6. Karimi, Hoda, Shahriar Mahdavi, Behnam Asgari Lajayer et al. “Insights on the bioremediation technologies for pesticide-contaminated soils.” *Environmental Geochemistry and Health* 44, no. 4 (2022): 1329–1354.
7. Kumar, Anil, Kapur Singh, Pushpendra Verma et al. “Effect of nitrogen and zinc nanofertilizer with the organic farming practices on cereal and oil seed crops.” *Scientific Reports* 12, no. 1 (2022): 1–7.
8. Kumar, Manish, Raghvendra Saxena, Rajesh Singh Tomar et al. “Role of methylotrophic bacteria in climate change mitigation.” *Microbes for Climate Resilient Agriculture* (2018): 149–164.
9. Cavicchioli, Ricardo, William J. Ripple, Kenneth N. Timmis et al. “Scientists’ warning to humanity: Microorganisms and climate change.” *Nature Reviews Microbiology* 17, no. 9 (2019): 569–586.
10. Dutta, Himangshu, and Angshu Dutta. “The microbial aspect of climate change.” *Energy, Ecology and Environment* 1, no. 4 (2016): 209–232.
11. Van Der Heijden, Marcel GA, Richard D. Bardgett, and Nico M. Van Straalen. “The unseen majority: Soil microbes as drivers of plant diversity and productivity in terrestrial ecosystems.” *Ecology Letters* 1, no. 3 (2008): 296–310.

12. Majumder, Nilanjana, and Koushik Biswas. "Soil microflora and its role in diminution of global climate change." In *Microbiomes and the Global Climate Change*, pp. 225–246. Springer, Singapore, 2021.
13. Wang, Jianlong, and Can Chen. "Biosorbents for heavy metals removal and their future." *Biotechnology Advances* 27, no. 2 (2009): 195–226.
14. Abatenh, Endeshaw, Birhanu Gizaw, Zerihun Tsegaye, and Misganaw Wassie. "The role of microorganisms in bioremediation: A review." *Open Journal of Environmental Biology* 2, no. 1 (2017): 038–046.
15. Lamichhane, Jay Ram, Silke Dachbrodt-Saaydeh, Per Kudsk, and Antoine Messéan. "Toward a reduced reliance on conventional pesticides in European agriculture." *Plant Disease* 100, no. 1 (2016): 10–24.
16. Hassanshahian, Mehdi, Mohammad Ahmadinejad, Hamid Tebyanian, and Ashraf Kariminik. "Isolation and characterization of alkane degrading bacteria from petroleum reservoir waste water in Iran (Kerman and Tehran provenances)." *Marine Pollution Bulletin* 73, no. 1 (2013): 300–305.
17. Singh, S. N., B. Kumari, and Shweta Mishra. "Microbial degradation of alkanes." In *Microbial Degradation of Xenobiotics*, pp. 439–469. Springer, Berlin, Heidelberg, 2012.
18. Phelan, Vanessa V., Wei-Ting Liu, Kit Pogliano, and Pieter C. Dorrestein. "Microbial metabolic exchange: The chemotype-to-phenotype link." *Nature Chemical Biology* 8, no. 1 (2012): 26–35.
19. Chauhan, S. K., N. Gupta, R. Walia et al. "Biomass and carbon sequestration potential of poplar-wheat inter-cropping system in irrigated agro-ecosystem in India." *Journal of Agricultural Science and Technology A* 1, no. 4 (2011): 575–586.
20. Dahuja, Anil, Ranjeet R. Kumar, Akshay Sakhare et al. "Role of ATP-binding cassette transporters in maintaining plant homeostasis under abiotic and biotic stresses." *Physiologia Plantarum* 171, no. 4 (2021): 785–801.
21. Paterson, Mark. *The Senses of Touch: Haptics, Affects and Technologies.* Berg, Oxford; Newyork, 2007.
22. Kabiraj, Ashutosh, Krishnendu Majhi, Urmi Halder et al. "Role of plant growth-promoting Rhizobacteria (PGPR) for crop stress management." In *Sustainable Agriculture in the Era of Climate Change*, pp. 367–389. Springer, Cham, 2020.
23. Yadav, Reena, Radhakrishnan Thankappan, and Abhay Kumar. "Novel approaches for genome editing to develop climate smart crops." In *Microbiomes and the Global Climate Change*, pp. 267–291. Springer, Singapore, 2021.
24. Osakabe, Yuriko, and Keishi Osakabe. "Genome editing with engineered nucleases in plants." *Plant and Cell Physiology* 56, no. 3 (2015): 389–400.
25. Fang, Yujie, and Lizhong Xiong. "General mechanisms of drought response and their application in drought resistance improvement in plants." *Cellular and Molecular Life Sciences* 72, no. 4 (2015): 673–689.
26. Li, Ting, Bo Liu, Martin H. Spalding et al. "High-efficiency TALEN-based gene editing produces disease-resistant rice." *Nature Biotechnology* 30, no. 5 (2012): 390–392.
27. Gruner, Katrin, Tobias Esser, Johanna Acevedo-Garcia et al. "Evidence for allele-specific levels of enhanced susceptibility of wheat mlo mutants to the hemibiotrophic fungal pathogen Magnaporthe oryzae pv. Triticum." *Genes* 11, no. 5 (2020): 517.
28. Metje-Sprink, Janina, Thorben Sprink, and Frank Hartung. "Genome-edited plants in the field." *Current Opinion in Biotechnology* 61 (2020): 1–6.
29. Haun, William, Andrew Coffman, Benjamin M. Clasen et al. "Improved soybean oil quality by targeted mutagenesis of the fatty acid desaturase 2 gene family." *Plant Biotechnology Journal* 12, no. 7 (2014): 934–940.

30. Jaganathan, Deepa, Karthikeyan Ramasamy, Gothandapani Sellamuthu et al. "CRISPR for crop improvement: An update review." *Frontiers in Plant Science* 9 (2018): 985.
31. Yin, Kangquan, and Jin-Long Qiu. "Genome editing for plant disease resistance: Applications and perspectives." *Philosophical Transactions of the Royal Society B* 374, no. 1767 (2019): 20180322.
32. Wang, Fujun, Chunlian Wang, Piqing Liu et al. "Enhanced rice blast resistance by CRISPR/Cas9-targeted mutagenesis of the ERF transcription factor gene OsERF922." *PloS one* 11, no. 4 (2016): e0154027.
33. Wang, Lijuan, Shanchun Chen, Aihong Peng et al. "CRISPR/Cas9-mediated editing of CsWRKY22 reduces susceptibility to *Xanthomonas citri* subsp. citri in Wanjincheng orange (Citrus sinensis (L.) Osbeck)." *Plant Biotechnology Reports* 13, no. 5 (2019): 501–510.
34. Surajudeen, Abdulsalam and Adaji Benjamin, Omale. "Comparison of biostimulation and bioaugmentation techniques for the remediation of used motor oil contaminated soil." *Brazilian Archives of Biology and Technology* 52 (2009): 747–754.
35. Zehra, Andleeb, Manish K. Dubey, Mukesh Meena et al. "Role of penicillium species in bioremediation processes." In *New and Future Developments in Microbial Biotechnology and Bioengineering*, 247–260. Elsevier, 2018.
36. Idris, Safiyanu, Abdulwahid Isah Adamu, Umar Suleiman Abubakar et al. "Review on comparative study on bioremediation for oil spills using microbes." *Research Journal of Pharmaceutical, Biological and Chemical Sciences (RJPBCS)* 6, no. 6 (2015): 783–790.
37. Aranda, Elisabet, René Ullrich, and Martin Hofrichter. "Conversion of polycyclic aromatic hydrocarbons, methyl naphthalenes and dibenzofuran by two fungal peroxygenase." *Biodegradation* 21, no. 2 (2010): 267–281.
38. Hesham, Abd El-Latif, Saad A. Alamri, and Sardar Khan. "Isolation and molecular genetic characterization of a yeast strain able to degrade petroleum polycyclic aromatic hydrocarbons." *African Journal of Biotechnology* 8, no.10 (2009): 2218–2223.

9 Impact of microbial biofilm community shifts on the degradation of microplastics in cold habitats

*Saheli Ghosh, Asifa Qureshi** *and Hemant J. Purohit*

CONTENTS

9.1 INTRODUCTION

Synthetic plastics are accumulating at an enormous rate when released into the environment. This is regarded as detrimental.[1–4] As plastic films are chemically inert, they possess great resistance to attack through physical, chemical

* Corresponding author e-mail: a_qureshi@neeri.res.in

DOI: 10.1201/9781003354031-9

and biological means.[5] Plastic waste that is dumped reaches the oceanic gyres, accounting for 50–80% of the pollution contaminating marine habitats globally.[6–11] According to Ioakeimidis et al.,[12] the bottom of the Black Sea and the Eastern Mediterranean Sea consists of plastics, accumulating up to 95% among other litter items. Furthermore, this debris becomes persistent and recalcitrant in the environment,[13] absorbing pollutants and organic matter.[14] It is weathered into microscopic chunks, which are taken up in the surface of the sea water, sediment and atmosphere as litter.[15] Polyethylene, low-density polyethylene (LDPE), high-density polyethylene (HDPE), polypropylene, etc. are the major sources of plastic litter.[13,16] Transportation of these recalcitrant polymers along the oceanic gyres has been raising concern over their degradation. Plastic films accumulate microbial flora in the form of biofilms, which significantly contribute to their degradation, as the extracellular matrix provides substrate bioavailability and increased cell viability.[17] Hence, microbial community shifts alter the degradation potential and buoyancy of plastic marine debris (PMD).[18] Assessment of this aspect of plastic remediation must be monitored. In the present chapter, we have described the effects of microbial community shifts on the degradation of various microplastic films. The reports illustrate that microbial community shifts significantly affect the degradation process. Furthermore, the effects of microbial community change and pathogen entrapment are also described, which contribute to the degradation process.

9.2 PLASTISPHERE ASSEMBLAGES IN COLD HABITATS

Microbes have colonised PMD since the 1970s, and the occurrence of diatoms has also been indicated.[19] In marine water, PMD constitutes a novel habitat that serves to harbour a plethora of microbial flora and fauna. Plastic has turned out to be a ubiquitous form of marine debris, which represents a significant and growing pollution issue globally. The northern shores of the Mediterranean Sea harbour plastics at concentrations of up to 10^5 pieces km^{-1},[7] equivalent to the numbers found in the 'great Pacific garbage patch' weighing over 0.25 trillion tonnes in the North Atlantic Subtropical Gyre.[8] The literature points out that plastic input into the oceans from land is estimated to have increased sharply in 10 years.[20] It serves as a niche that allows the assemblage and enrichment of specific microbes, offering support for their growth and decreased predation.[8] However, available reports of the PMD analysis are inadequate. Microbial communities adherent in PMDs, called the *plastisphere*, have been studied.[21–24] This showed an increased diversity of autotrophs, heterotrophs, predators and symbionts, followed by pathogens, which could degrade either the polymer or the adsorbed pollutants.[25,26] Microbial biofilms can be formed on the plastic surface, which become visible in a week and make it lighter by modifying its physical properties. The hydrophilicity was increased and the polymer became neutrally buoyant from positive buoyancy.[27] Biofouling increases the density of PMDs, which allows them to sink to the sea floor.[28–30] In this way, fouled debris could attract invertebrates, which could graze upon them.[31–38] Available reports point out the ill effects of microplastics on the indigenous biota, which may involve physical blockage due to the consumption of PMDs as food, including their capability to transport algae, pathogens and harmful compounds.[32,33,38] Microplastic mediated communities' influences their impact on distribution and fate. Biofilms allow nutrient cycling and carbon transformation, which contributes to the formation of food webs and aids in ecosystem functioning in streams and other habitats.[13,39]

9.3 BIOFILM COMMUNITY DIVERSITY IN THE PLASTISPHERE

In marine environments, plastic-associated microorganisms are taxonomically different from the surrounding sea waters.[24,40] Oberbeckmann et al.[30] characterised the plastic communities in polyethylene terephthalate (PET) water bottles, glass and North Pacific sea water. It was found that the microbial community was nearly the same in PET and glass bottles; however, it significantly differed in the surrounding waters, indicating the enrichment of the bacteria in the PMDs.[29] Similarly, McCormick et al. (2016) compared the genera selected on plastic debris with the surrounding sea water and waste-water treatment plant (WWTP) outfall. They included *Arcobacter, Aeromonas, Pseudomonas, Aquabacterium* and *Zymophilus*. These organisms are mainly reported for plastic degradation as well as pathogenesis. *Aquabacterium commune* is commonly found in the biofilm communities of drinking water,[27,41] and colonisation of *Arcobacter* sp. in LDPE was found on sediments of seawaters and oceans. This report also points out the emergence of microplastic-colonising *Arcobacter* sp. in freshwater habitats.[28] It indicates that certain taxa survive on polymers when they are transported. Since the habitats are dissimilar, the adherent biofilm community has relatively higher microbial diversity in seston than in deeper sea waters. However, the presence of higher eukaryotes, as well as the effects of habitats on the colonisation of the microbial communities, affects the plastic degradation. Eich et al.[10] analysed the microbial community in PMDs and pointed out that biofilm formation was increased in polyethylene films in both benthic and pelagic zones. On the contrary, diatom diversity differed in PE films of both the zones. This may be due to the different surface properties, light intensity and shear forces. Those bacterial species that were present in the pelagic zone were reported to be absent in the benthic zones, and vice versa. However, the presence of *Bacillus* sp. in PMDs cannot be ignored, as they are also an important determinant in plastic degradation. When plastic items were collected from surface of the North Pacific Gyre and analysed using scanning electron microscopy, pennate diatoms (1097 ± 154 mm) and *Bacillus* sp. (mean 1664 ± 247 cells mm^{-2}) were found to be most abundant, along with cocci, dinoflagellates, centric diatoms, radiolarians and coccolithophores. Bacterial abundance in patches was found on foamed polystyrene. The diatom abundance was more patchy and found on areas of greater thickness. Diversity in morphology was increased on larger-sized debris, and the pennate diatom community differed in structure.[10,42]

9.4 FACTORS AFFECTING MICROBIAL COMMUNITY SHIFTS IN THE PLASTISPHERE

There are a number of factors affecting microbial community shifts in the plastisphere: pollutant transportation and presence of pathogenic organisms, etc.[29,43]

9.4.1 Pollutant Transportation and Biodegradation

On the marine surface, some plastic debris gets heavier and tends to settle down more than the less heavy, buoyant debris. This has the tendency to accumulate on the sea surface, gets transported by oceanic currents and winds over long distances,it floats on the water surfaces.[5] Among the different classes of plastics, polyethylene, polypropylene, etc. have a specific gravity lower than that of marine waters and are thus positively buoyant, whereas other plastic films, such as polyvinyl chloride, are denser and have a tendency to settle down in the water column. However, fouling of plastic litter may convert positively buoyant debris to settle down in the deep sea on a scale from weeks to months.[10,42] When exposed to radiation such as UV, this debris gets partially degraded to short-chain compounds. These are further degraded chemically along with the interference of microbes, such as biodegradation. However, during degradation, fragments and nano-size particles are generated .[29,43]

9.4.2 Plastisphere Pathogenic Community and Its Effects

Microplastics or PMDs are more susceptible to adsorbing waterborne contaminants than macroplastics due to their higher surface area to volume ratio.[43] Available studies have pointed out that plastic acts as a vector for the spread of pathogenic microbes[22] Analysis of gene sequences of the adherent biofilm communities of PMDs allows the detection of transportation of pathogenically active microbes along with the debris. Among them, *Vibrio* sp. was detected on polypropylene, but to a lesser extent in polyethylene, in North Atlantic waters.[22] However, Schmidt et al.[34] detected several pathogenic species of *Vibrio* sp. affecting animals, corals and other higher organisms from mixed marine plastic samples.[34] Metagenomic analysis revealed the abundance of Type III and Type IV secretion systems in the PMDs.[6] However, genes related to horizontal gene transfer, coaggregation, and biofilm-formation were abundantly observed in the plastic films.[32] In addition to bacteria, plastic debris also transports eukaryotic microorganisms.[34] Harmful species of algae have been transported along the shores of the Mediterranean Sea.[6] Sequences related to *Pseudomonas* sp. were relatively abundant compared with *Aeromonas* sp. on all kinds of plastics in rivers, implying that such species used microplastics as a vector in pathogenic species spread. However, PET harbours *Tebnacibaculum* and *Aeromonas* sp.[3]

9.5 CHALLENGES AND GAPS IN PLASTIC RESEARCH

In literature, a significant gap exists when it comes to inferring the exact types of the microbes adherent to the plastic. The regulation of biofilm–plastic interactions needs to be studied. These studies need to ascertain the role of associated microflora in conferring the buoyancy of plastic films, thereby influencing the microbial community structure. Additionally, further research needs to be conducted on the entrapment of pathogenic microbes in the biofilm community. Oberbeckmann and coworkers[30] reported that the plastic debris carries pathogens from the faecal matter of marine organisms and humans. Coherent studies need to be conducted to ascertain the impacts of microorganisms of low trophic levels.[9,35] The impacts of plastisphere organisms of nutrient cycling and other geochemical processes must be investigated. The impact of adherent microflora on geochemical cycling processes, as well as interaction with the higher organisms in the sea waters, also affects their weathering and transportation. They may become transported from faecal matter.[1,9] The gut bacteria of mealworms, along with other microbial flora, can modify the topology and the surface characteristics of microplastics by virtue of degradation.[23] Hence, the biofilm–plastic interactions and the interference of higher organisms towards the plastic degradation need to be investigated.

9.6 DEGRADABILITY OF PLASTICS WITH RELEVANCE TO THE ADHERENT MICROBIAL COMMUNITY

Adherent microorganisms secrete extracellular polymeric substances on the plastic debris and form a biofilm. They utilise plastic either as a support or as a substrate. However, the diversity of microbial flora on the debris depends upon the type of plastic and the surrounding waters. Plastics are degraded physically by the action of UV light into shorter chains, which reduces their buoyancy and allows conditioning of the film.[1] Microbial flora then colonises over time in the plastic film and carries out further degradation processes, as shown in

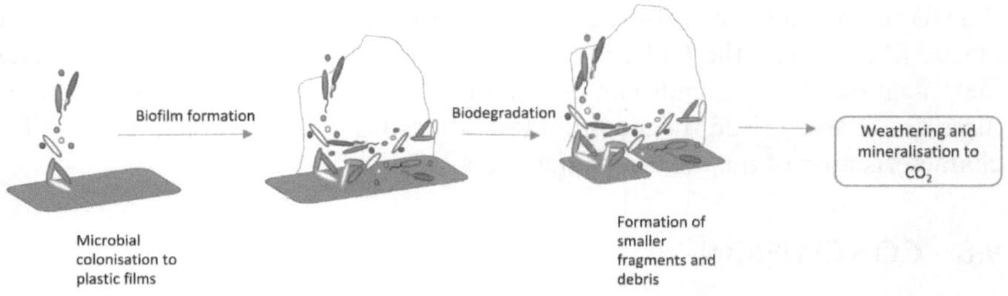

FIGURE 9.1 Biofilm formation of the microbial community and degradation of plastic films.

Figure 9.1. During the degradation process, extracellular enzymes are secreted by the underlying cells, further cleaving short-chain oligomers into dimers and monomers. Enzymes are secreted in the periplasmic space or the cell membrane and are transported through the cell membrane.[2,36] In some cases, the oligomers or dimers are further metabolised by the central degradation pathways to form the tricarboxylic acid (TCA) cycle intermediates.[23] As reported by Kawai et al. (2004), some oligomers can be directly internalised using biosurfactants synthesised by the respective microbes, thereby entering beta-oxidation.[19,20] On the contrary, these monomers could also enter the TCA cycle through the beta-oxidation pathway.[1]

9.7 ADVANCES IN PLASTISPHERE RESEARCH

Many analytical and omics tools are currently being used in plastisphere research. However, combinations of tools are applied in deciphering the microbial community adherent in PMDs. Several tools, such as metabarcoding, were developed for analysing protists, fungi and diatoms. However, this approach mainly targets some groups of microbes, not their metabolic pathways. To gain a bigger picture, meta-transcriptomics, microarrays or metagenomics could be used, with the former revealing the metabolic ability while metagenomics annotates the genetic machinery involved.[24] The existence of pathogens colonising PMDs is characterised by whole genome sequencing (WGS) followed by comparison of genomes and single nucleotide polymorphisms. This enables certain genes to be gained through horizontal gene transfer mechanisms, thereby acquiring pathogenicity and degradation capacity.[19] Further advancements allows to gather information regarding the different stages of biofilm on plastic debris. Stable isotope labelling is currently being employed to analyse degradation properties at the single-cell level. Studies involving coupling of heavy water labelling with Raman spectroscopy and cell sorters enables the identification of metabolically dynamic taxa *in situ*. Raman spectroscopy, when integrated with FISH (fluorescence in situ hybridisation), may be employed to decipher the occurrence of a particular taxon.[41] Fourier transform infrared (FTIR) techniques have also been applied to detect degraded by-products of plastic films. When the FTIR is combined with mass spectrometers, the exact identification of compounds can be obtained. The chemical constituents of biofilms are characterised, hence providing a feasible and convenient method for characterisation of the adherent microflora.[42,43]

9.8 CONCLUSION

Microplastics predominate in the form of litter in marine habitats, which are run-offs from anthropogenic sources. Adherent biofilm communities are responsible for their degradation in these cold habitats. Transportation of this debris through oceanic currents allows it to spread. However, microbes adhering to plastic films degrade them, which in turn escalates biofilm formation, availability and sharing of substrates, and maintains cell viability in the underlying cells. Reports are available on various physical, chemical and biodegradation processes employed. The available reports indicate the usage of single species as well as community biofilms. The current chapter describes the biofilm community change towards the biodegradation of microplastics present in cold marine habitats. Furthermore, microbial communities during the course of degradation acquire new traits as well as harbouring potential pathogens. Many combinations of tools can be used to decipher these traits of the microbial community harboured in plastic litter, enhancing the understanding of the microbial flora, which drives the degradation process and eventually aids in plastic waste remediation.

ACKNOWLEDGEMENT

All the authors are thankful to CSIR-NEERI for constant support and inspiration and providing infrastructural facilities (CSIR-NEERI/KRC/2020/JAN/EBGD/1). Funds from DBT (BT/PR16149/NER/95/85/2015) projects are acknowledged.

REFERENCES

1. Afolabi, T. A., and O. O. Ogundiran. "Mechanical and biodegradability properties of hydroxypropyl and cross-linked starch-Low density polyethylene (LDPE) composite." *Journal of Chemical Society of Nigeria* 44, no. 3 (2019).
2. Albertsson, A-C., and Z. G. Banhidi. "Microbial and oxidative effects in degradation of polyethene." *Journal of Applied Polymer Science* 25, no. 8 (1980): 1655–1671.
3. Anjana, K., M. Hinduja, K. Sujitha, and G. Dharani. "Review on plastic wastes in marine environment–Biodegradation and biotechnological solutions." *Marine Pollution Bulletin* 150 (2020): 110733.
4. Barnes, David KA, Francois Galgani, Richard C. Thompson, and Morton Barlaz. "Accumulation and fragmentation of plastic debris in global environments." *Philosophical Transactions of the Royal Society B: Biological Sciences* 364, no. 1526 (2009): 1985–1998.
5. Beaumont, Nicola J., Margrethe Aanesen, Melanie C. Austen et al. "Global ecological, social and economic impacts of marine plastic." *Marine Pollution Bulletin* 142 (2019): 189–195.
6. Bryant, Jessica A., Tara M. Clemente, Donn A. Viviani et al. "Diversity and activity of communities inhabiting plastic debris in the North Pacific Gyre." *MSystems* 1, no. 3 (2016): e00024-16.
7. Carpenter, Edward J., and K. L. Smith Jr. "Plastics on the Sargasso sea surface." *Science* 175, no. 4027 (1972): 1240–1241.
8. Collignon, Amandine, Jean-Henri Hecq, François Galgani et al. "Annual variation in neustonic micro-and meso-plastic particles and zooplankton in the Bay of Calvi (Mediterranean–Corsica)." *Marine Pollution Bulletin* 79, no. 1–2 (2014): 293–298.
9. Diaz, Robert J., Martin Solan, and Raymond M. Valente. "A review of approaches for classifying benthic habitats and evaluating habitat quality." *Journal of Environmental Management* 73, no. 3 (2004): 165–181.
10. Eich, Andreas, Tobias Mildenberger, Christian Laforsch, and Miriam Weber. "Biofilm and diatom succession on polyethylene (PE) and biodegradable plastic

bags in two marine habitats: Early signs of degradation in the pelagic and benthic zone?." *PloS One* 10, no. 9 (2015): e0137201.
11. Fotopoulou, Kalliopi N., and Hrissi K. Karapanagioti. "Degradation of various plastics in the environment." *Hazardous Chemicals Associated with Plastics in The Marine Environment* (2017): 71–92.
12. Ioakeimidis, Christos, K. N. Fotopoulou, H. K. Karapanagioti et al. "The degradation potential of PET bottles in the marine environment: An ATR-FTIR based approach." *Scientific Reports* 6, no. 1 (2016): 1–8.
13. Gewert, Berit, Merle M. Plassmann, and Matthew MacLeod. "Pathways for degradation of plastic polymers floating in the marine environment." *Environmental Science: Processes & Impacts* 17, no. 9 (2015): 1513–1521.
14. Ghosh, Saheli, Asifa Qureshi, and Hemant J. Purohit. "Aromatic compounds and biofilms: Regulation and interlinking of metabolic pathways in bacteria." In *Microbial Metabolism of Xenobiotic Compounds*, pp. 145–164. Springer, Singapore, 2019.
15. Ghosh, Saheli, Asifa Qureshi, and Hemant J. Purohit. "Enhanced expression of catechol 1, 2 dioxygenase gene in biofilm forming *Pseudomonas mendocina* EGD-AQ5 under increasing benzoate stress." *International Biodeterioration & Biodegradation* 118 (2017): 57–65.
16. Ghosh, Saheli, Asifa Qureshi, and Hemant J. Purohit. "Microbial degradation of plastics: Biofilms and degradation pathways." *Contaminants in Agriculture and Environment: Health Risks and Remediation* 1 (2019): 184–199.
17. Hirai, Hisashi, Hideshige Takada, Yuko Ogata et al. "Organic micropollutants in marine plastics debris from the open ocean and remote and urban beaches." *Marine Pollution Bulletin* 62, no. 8 (2011): 1683–1692.
18. Karapanagioti, Hrissi K., Satoshi Endo, Yuko Ogata, and Hideshige Takada. "Diffuse pollution by persistent organic pollutants as measured in plastic pellets sampled from various beaches in Greece." *Marine Pollution Bulletin* 62, no. 2 (2011): 312–317.
19. Kawai, Fusako, Masaji Watanabe, Masaru Shibata et al. "Experimental analysis and numerical simulation for biodegradability of polyethylene." *Polymer Degradation and Stability* 76, no. 1 (2002): 129–135.
20. Kawai, Fusako, Masaji Watanabe, Masaru Shibata et al. "Comparative study on biodegradability of polyethylene wax by bacteria and fungi." *Polymer Degradation and Stability* 86, no. 1 (2004): 105–114.
21. Keswani, Anisha, David M. Oliver, Tony Gutierrez, and Richard S. Quilliam. "Microbial hitchhikers on marine plastic debris: Human exposure risks at bathing waters and beach environments." *Marine Environmental Research* 118 (2016): 10–19.
22. Kirstein, Inga V., Sidika Kirmizi, Antje Wichels et al. "Dangerous hitchhikers? Evidence for potentially pathogenic *Vibrio* spp. on microplastic particles." *Marine Environmental Research* 120 (2016): 1–8.
23. Koutny, Marek, Jacques Lemaire, and Anne-Marie Delort. "Biodegradation of polyethylene films with prooxidant additives." *Chemosphere* 64, no. 8 (2006): 1243–1252.
24. Law, Kara Lavender, Skye Morét-Ferguson, Nikolai A. Maximenko et al. "Plastic accumulation in the North Atlantic subtropical gyre." *Science* 329, no. 5996 (2010): 1185–1188.
25. Lobelle, Delphine, and Michael Cunliffe. "Early microbial biofilm formation on marine plastic debris." *Marine Pollution Bulletin* 62, no. 1 (2011): 197–200.

26. Zettler, Erik R., Tracy J. Mincer, and Linda A. Amaral-Zettler. "Life in the 'plastisphere': Microbial communities on plastic marine debris." *Environmental Science & Technology* 47, no. 13 (2013): 7137–7146.
27. McCormick, Amanda R., and Timothy J. Hoellein. "Anthropogenic litter is abundant, diverse, and mobile in urban rivers: Insights from cross-ecosystem analyses using ecosystem and community ecology tools." *Limnology and Oceanography* 61, no. 5 (2016): 1718–1734.
28. Mincer, Tracy J., Erik R. Zettler, and Linda A. Amaral-Zettler. "Biofilms on plastic debris and their influence on marine nutrient cycling, productivity, and hazardous chemical mobility." *Hazardous Chemicals Associated with Plastics in the Marine Environment* (2016): 221–233.
29. Neto, José Antonio Baptista, Christine Gaylarde, Iwona Beech et al. "Microplastics and attached microorganisms in sediments of the Vitória bay estuarine system in SE Brazil." *Ocean & Coastal Management* 169 (2019): 247–253.
30. Oberbeckmann, Sonja, Martin GJ Löder, and Matthias Labrenz. "Marine microplastic-associated biofilms: a review." *Environmental Chemistry* 12, no. 5 (2015): 551–562.
31. Ojeda, Jesus J., Maria E. Romero-Gonzalez, and Steven A. Banwart. "Analysis of bacteria on steel surfaces using reflectance micro-Fourier transform infrared spectroscopy." *Analytical Chemistry* 81, no. 15 (2009): 6467–6473.
32. Parthasarathy, Anutthaman, Anna Christina Tyler, Matthew J. Hoffman et al. "Is plastic pollution in aquatic and terrestrial environments a driver for the transmission of pathogens and the evolution of antibiotic resistance?" *Environmental Science & Technology* 53, no.4 (2019): 1744–1745.
33. Reisser, Julia, Jeremy Shaw, Gustaaf Hallegraeff et al. "Millimeter-sized marine plastics: A new pelagic habitat for microorganisms and invertebrates." *PloS One* 9, no. 6 (2014): e100289.
34. Schmidt, Victor T., Julie Reveillaud, Erik Zettler et al. "Oligotyping reveals community level habitat selection within the genus Vibrio." *Frontiers in Microbiology* 5 (2014): 563.
35. Schratzberger, Michaela, J. M. Gee, H. L. Rees et al. "The structure and taxonomic composition of sublittoral meiofauna assemblages as an indicator of the status of marine environments." *Journal of the Marine Biological Association of the United Kingdom* 80, no. 6 (2000): 969–980.
36. Shah, Aamer Ali, Fariha Hasan, Abdul Hameed, and Safia Ahmed. "Biological degradation of plastics: A comprehensive review." *Biotechnology Advances* 26, no. 3 (2008): 246–265.
37. Sudhakar, M., Mukesh Doble, P. Sriyutha Murthy, and Ramasamy Venkatesan. "Marine microbe-mediated biodegradation of low-and high-density polyethylenes." *International Biodeterioration & Biodegradation* 61, no. 3 (2008): 203–213.
38. Urbanek, Aneta K., Waldemar Rymowicz, and Aleksandra M. Mirończuk. "Degradation of plastics and plastic-degrading bacteria in cold marine habitats." *Applied Microbiology and Biotechnology* 102, no. 18 (2018): 7669–7678.
39. Van Franeker, Jan A., and Kara Lavender Law. "Seabirds, gyres and global trends in plastic pollution." *Environmental Pollution* 203 (2015): 89–96.
40. Wagner, Martin, Christian Scherer, Diana Alvarez-Muñoz et al. "Microplastics in freshwater ecosystems: What we know and what we need to know." *Environmental Sciences Europe* 26, no. 1 (2014): 1–9.
41. Wang, Yun, Wei E. Huang, Li Cui, and Michael Wagner. "Single cell stable isotope probing in microbiology using Raman microspectroscopy." *Current Opinion in Biotechnology* 41 (2016): 34–42.

42. Woodall, Lucy C., Anna Sanchez-Vidal, Miquel Canals et al. "The deep sea is a major sink for microplastic debris." *Royal Society Open Science* 1, no. 4 (2014): 140317.
43. Xu, Xiyuan, Shuai Wang, Fenglei Gao et al. "Marine microplastic-associated bacterial community succession in response to geography, exposure time, and plastic type in China's coastal seawaters." *Marine Pollution Bulletin* 145 (2019): 278–286.

10 A review on the bioprospection of psychrophilic bacteria and their potential applications for sustainability

Neeraja P. Dhole#, Mudasir A. Dar#, Kalim Ullah, Rongrong Xie, Jianzhong Sun, Mohd. Shahnawaz and Radhakrishna S. Pandit**

CONTENTS

* Corresponding authors e-mail: mudasir.dar@unipune.ac.in; rspandit@unipune.ac.in
# Authors contributed equally and are both first authors

DOI: 10.1201/9781003354031-10

10.1 INTRODUCTION

The world's oceans occupy 71% of the Earth's surface, having a temperature below 5 °C. About 20% of the Earth's surface is composed of cold environments such as glaciers, ice sheets, permafrost, polar sea ice and snow.[1] Due to this, 80% of the Earth has a temperature below 5 °C.[2] Further, planet Earth harbours life in almost every type of habitat. For example, some organisms have adapted themselves to survive in the harshest environments having extreme pH, temperature and salinity. Among these features, temperature tolerance is an important adaptation, as it affects all the biochemical processes required for survival. To this end, cold-adapted microorganisms can be regarded as the most successful colonisers of the planet Earth.[3] It has been suggested that microbial life on Earth may have originated in the subsurface of Martian permafrost or in warm diapirs of Europe's ice, at very low temperatures.[4] Among the cold-loving organisms, bacteria comprise a special group from the viewpoint of biotechnology as well as biology. To date, many extremophiles, particularly bacteria, fungi and algae, have been described that can survive at low temperatures. The lower limit of growth of some unicellular bacteria is reported to be –20 °C, whereas multicellular organisms stop growing at –2 °C.[5]

The cold-loving bacteria, also known as psychrophiles, are a significant part of the biosphere, as they have become established at the poles, mountains, deep oceans (where they make up 95% of the total biomass), sea ice, ice packs, etc. Recently, the presence of psychrophilic bacteria in space has also been reported. Table 10.1 lists some examples of cold-adapted bacteria. Among these bacteria, some can grow at below-freezing temperatures, showing active metabolism at temperatures as low as –20 °C.[6] These adaptations in cold-tolerant microorganisms make them a perfect tool for biotechnology.[7] In view of this objective, here we have discussed the recent trends in the bioprospection of psychrophilic bacteria. Further, a detailed account of the potential applications of these promising candidates in biotechnology is also elaborated.

10.2 PSYCHROPHILES OR PSYCHROTROPHS?

In literature, the terms *psychrophiles* and *psychrotrophs* are arguably interchangeable. Psychrophiles *sensu stricto*, or cold lovers, are organisms having

TABLE 10.1
Morphology and phylogenetic affiliation of some commonly reported cold-adapted bacteria

Sr. No.	Bacteria	Gram's stain	Morphology	Phylum
1	*Flavobacterium petrolei*	–ve	Rods without flagella	Bacteroidetes[8]
2	*Sphingopyxis chilensis*	–ve	Motile rods	Proteobacteria[9]
3	*Aquaspirillum arcticum*	–ve	Weakly motile rods	Proteobacteria[10]
4	*Arthrobacter flavus*	+ve	Rods, cocci, non-motile	Actinobacteria[11]
5	*Arthrobacter cryoconite*	+ve	Rods, cocci, non-motile	Actinobacteria[12]
6	*A. psychrophenolicus*	+ve	Rods, cocci, non-motile	Actinobacteria[13]
7	*Bacillus marinus*	–ve	Motile rods	Firmicutes[14]
8	*Colwellia beringensis*	–ve	Motile rods	Proteobacteria[15]
9	*Octadecabacter arcticus* gen., *O. antarcticus*	–ve	Non-motile rods	Proteobacteria[16]

growth optima at 15 °C or below, the highest temperature limit at 20 °C, and the lower temperature limit at 0 °C or below. In contrast, psychrotrophs are organisms that, although they can survive at 0 °C, have an optimum growth temperature of 20–25 °C. In other words, they can be referred to as cold-tolerant mesophilic organisms.[17] Generally, psychrotrophs are much more abundant as compared with psychrophilic organisms, due to their diversity and extended lag phase. They are more adaptable and diverse, as they can thrive in an extensive range of temperatures.[18] The lowest temperature limit for psychrophiles at which they can reproduce is defined to be –12 °C, while for metabolism it is –20 °C.[19] However, cellular functioning has been recently proposed to continue at temperatures lower than –20 °C.[20]

10.3 LOW TEMPERATURE: A CHALLENGE FOR SURVIVAL

The internal temperature of microbes is similar to the temperature outside. There is no way to control the temperature, unlike other stresses such as pH and osmolarity.[21] A temperature below optimum slows the growth of the organism. In general, the effects of cold shock are not well defined, as they are for heat shock. However, transient arrest of growth occurs at low temperatures, causing the production of several special proteins called cold shock proteins (CSPs).[22] These CSPs affect the overall metabolism of the cell. Cold stress alters not only metabolic functions but a variety of structures as well. In other words, in cold habitats, the microbes show a variety of metabolic adaptations/alterations, some of which are elaborated as follows.

10.3.1 Lag phase

The normal temperature range of the cell is an important parameter for the survival and growth of organisms because it has a direct effect on the length of the lag phase.[23] However, an extended lag phase can occur in mesophilic microorganisms upon exposure to low temperatures. One such example is *Escherichia coli*, where the lag phase is observed in cells growing at or below 8 °C. An extended lag phase is also reported in the bacterium *Listeria monocytogenes* at low temperatures such as 4 °C.[24]

10.3.2 Plasma membrane

The barrier function of the plasma membrane varies at a lower temperature, as it depends upon the physical state of the membrane, which is quasi-fluid at normal temperatures.[2] Under cold stress, the fluid plasma membrane turns into a non-fluid array of fatty acyl chains, where membrane phospholipids contain more unsaturated fatty acids. This maintenance property of fatty acids is called thermal regulation of fatty acids.[25] Moreover, the saturated fatty acid laurate is replaced by the appearance of an unsaturated fatty acid, palmitoleate.[26] In cold habitats, plasma membrane functions such as active transport and protein secretion are decreased.[22] At low temperature, microbes require a higher volume of organic substrates, exaggerating the effect of nutrient limitations. Hence, the permeability function of the plasma membrane increases at low temperature.[27]

10.3.3 Modification of nucleic acids and macromolecules

The stabilisation of secondary structures of nucleic acids occurs under low temperatures. The principal proteins of cold shock, such as Csp family proteins, are involved in preventing the formation of secondary structures of nucleic acids. The Csps are also involved in the destruction of constructed RNA. As a result of cold shock, DNA becomes more and more negatively supercoiled, and this leads to the transcription of cold-induced genes.[28] Enzymes like DNA topoisomerase, DNA gyrase and histone-like HU proteins are the key players in prevention of transcription of proteins other than Csps at low temperatures. Moreover, cold temperatures prevent translation of proteins within the cell except for the cold shock protein family. However, some major changes in the conformation, folding and flexibility of macromolecules are also noted due to cold exposure within the cells.[29]

10.3.4 Ribosome alterations

The extremophiles require a higher concentration of ribosomes than microorganisms that live at ordinary room temperature. This adaptation may be attributed to the translational block induced by the cold stress. Further, at low temperatures, there is a significant decrease in the functional polysomes as well as peptide elongation rates.[3] The ribosomal miscoding rate is much lower in psychrophiles as compared with mesophiles at similar temperatures.[30] The optimum growth temperature of a microorganism is directly related to the heat stability of ribosomes.[31]

10.3.5 Pigment production and other changes

It has been shown that the production of pigments at low temperatures is better than at normal temperatures.[32] Several other cases, such as brilliant yellow pigment production by *Flavobacterium* at 5 °C[33] and blue-black pigment formation by *Pseudoalteromonas nigrifaciens* at 4 °C[34] and *Pseudomonas* species at 3 °C[35] have been studied. In *Flavobacterium frigidarium*, the yellow pigment production is attributed to β-carotene-3,3-H,diol.[36] Recently, several researchers have observed a high frequency of pigment production in psychrophilic bacteria such as those isolated from ice cores[37] or sea waters[38] and glaciers.[39] These inferences suggest that the pigmentation mechanism is involved in the adaptation of bacteria to cold environments. In the Antarctic psychrotolerant bacterium *Sphingobacterium antarcticus*, the carotenoid pigments cause modulation of membrane fluidity,[40] thereby augmenting bacterial survival under cold stress. Dieser and coworkers[41] also found that carotenoid pigments provide resistance to environmental stresses in Antarctic heterotrophic bacteria. However, the mechanism relies on several factors, including pigment structure, location and pigment concentration in the cells. The carotenoid pigmented strains are also found to be more resistant to freeze-thaw cycles as compared with non-pigmented counter strains.

While most of the enzymes are inactivated or inhibited at cold temperatures, it is seen that enzyme production is induced at low temperature for certain enzymes. Production of proteinase and lipase is induced at low temperatures in *Pseudomonas* spp.[37] Also, cold acclimation associated sequences, selective capture of transcribed sequences (SCOTS), were found to be expressed more at low temperatures (10 °C) than at higher temperatures (37 °C) in the food pathogen *Listeria monocytogenes*[42] (Figure 10.1).

10.4 APPLICATIONS OF COLD-LIVING BACTERIA

As biotechnology is a natural by-product of biodiversity, psychrophilic organisms are very important from the industrial viewpoint (Figure 10.2). Psychrophiles possess a vast array of industrial significance. Broadly, the applications of psychrophiles can be given under the following categories.

10.4.1 Cold-adapted enzymes

The tremendous advantages offered by psychrophilic enzymes (Table 10.2), such as higher specific activities, high specificity and stability, high k_{cat} value

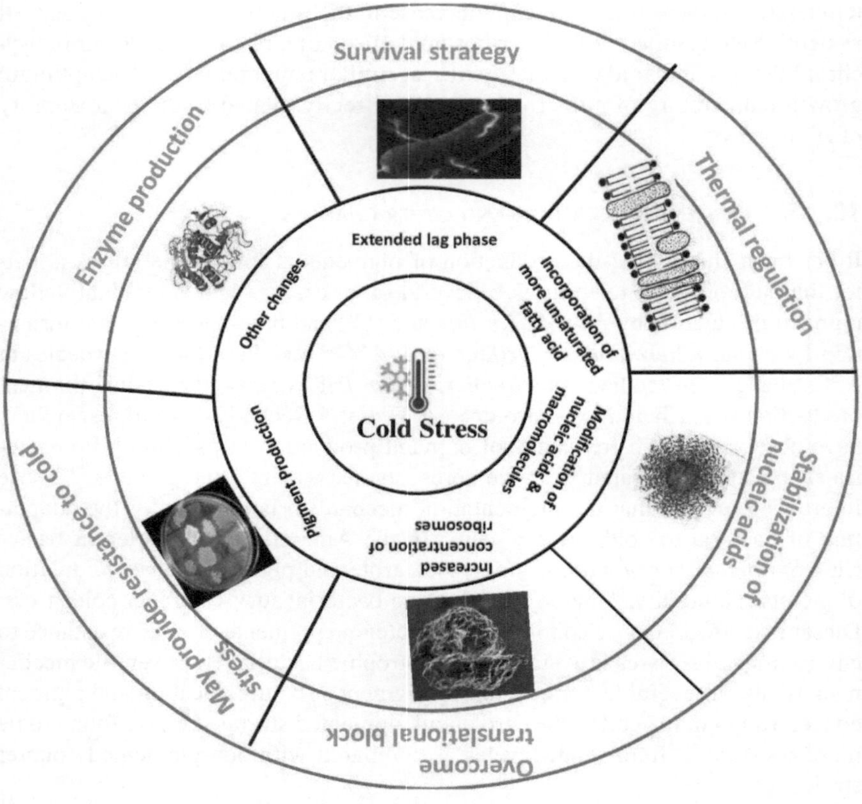

FIGURE 10.1 Overview of the different survival strategies employed by psychrophilic bacteria to withstand harsh environments.

at low temperatures, thermal stability and functioning in organic solvents, makes them suitable for industrial applications.[43] In addition, the biochemical reactions are energy efficient due to reduced reaction temperature, avoid unwanted bacterial contamination, and can be rapidly inactivated at high temperatures, which are vital for the food industry.[39] In the future, psychrophilic enzymes may likely be of higher importance than thermophilic enzymes due to their cost-efficient nature. Psychrophilic enzymes produced by genera like *Acinetobacter*, *Aquaspirillium*, *Arthrobacter*, *Bacillus*, *Carnobacterium*, *Clostridium*, *Cytophaga*, *Flavobacterium*, *Marinomonas*, *Moraxella*, *Moritella*, *Paenibacillus*, *Planococcus*, *Pseudoalteromonas*, *Pseudomonas*, *Psychrobacter*, *Shewanella*, *Vibrio* and *Xanthomonas* have already been proved important for industrial applications.[44]

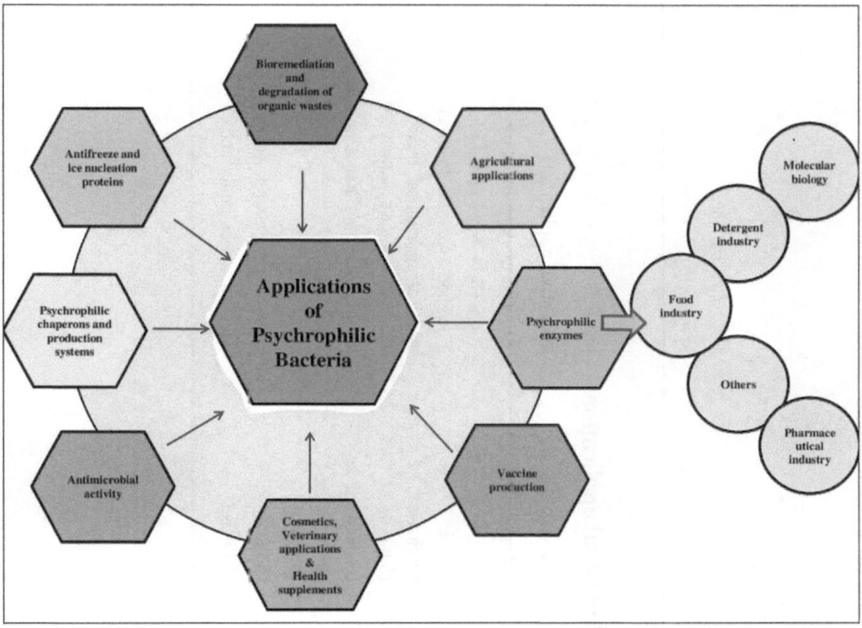

FIGURE 10.2 Industrial applications of psychrophilic or cold-tolerant bacteria and their enzymes.

10.4.2 Food Industry

The use of psychrophilic enzymes is common in the food industry, as the food material is generally treated at ambient temperature to prevent any possible spoilage, loss of nutritional value, and taste change due to heat-sensitive reactants and products.[54] Lactose intolerance is a problem faced by two-thirds of the world's population. Hence, lactose obtained from milk is hydrolysed to its constituents, glucose and galactose, using β-galactosidase from *Kluyveromyces* spp. at 30–40 °C. Unfortunately, this process faces the problem of bacterial contamination due to a long period of processing. However, if an alternative psychrophilic β-galactosidase were used in the procedure, it would result in 70–80% lactose hydrolysis with a lower risk of bacterial contamination. Spoilage of heat-sensitive materials used in food industries can be avoided by the use of psychrophilic lipases.[43,54] The use of lipases obtained from cold-active *Pseudomonas* spp. is very important in the dairy and juice industries.[55] Apart from glycosidases, psychrophilic proteases are used in the fermentation of soy sauce and fish without affecting their nutritional value. They are used for tenderisation and taste enhancement of meat products.[56] Moreover, the cold-active proteases are potential components

TABLE 10.2
Bacterial sources of the cold-tolerant enzymes and their potential industrial applications

Sr. No.	Bacteria	Source	Enzyme	Industrial application
1	*Paenibacillus* sp. R4	Soil	Esterase	Biomedical bioremediation[45]
2	*Rhodococcus* sp.	Atlantic hagfish, *Myxine glutinosa*	Esterase	Biomedical, bioremediation, catalysis[46]
3	*Pseudomonas* sp., *LSK25*	Soil	Lipase	Food additive, water treatment, laundry[47]
4	*Arthrobacter psychrochitiniphilus* sp. nov.	Guano of penguins	Chitinase	Medical, agricultural, food[48]
5	*Pseudoalteromonas* sp. DY3	Sea sediments	Cellulase	Biofuels, Laundry, pulp[49]
6	*Paenibacillus* sp. C7	Mud and water	β-Glucosidases	Biofuel, food, surfactant, agricultural[50]
7	*Mrakia frigida*	European Alps	Pectate lyase	Food industry[39]
8	*Pseudoalteromonas haloplanktis* ANT/505	Marine	Pectate lyase	Food industry[51]
9	*Alteromonas* sp. O-7	Marine	Chitinase	Medical, agricultural, food[52]
10	*Arthrobacter* sp.	Antarctic waters	Chitinase	Medical, agricultural, food[53]

of pet foods to increase taste.[57] Psychrophilic feruloyl esterases also show multiple applications in the food industries.[58] Some enzymes, like amylases, proteases and xylanases, procured from psychrophilic bacteria could be utilised for the baking of dough by acting on starch, gluten and hemicellulases to reduce intrinsic enzyme activities.[54] Pectinases are very important enzymes in the food industries, which individually make up 40% of all food enzymes.[59] Similarly, psychrophilic pectinases are used for clearing fruit vegetable juices, for vinification and for extraction of oils.[60] Many companies, such as Novozyme and Genencor, use psychrophilic bacterial enzymes, i.e., proteases, amylases, xylanases, cellulases and lipases, in commercial products, including food and detergents.[61]

10.4.3 Laundry and Cleaning Processes

Currently, 30–40% of commercial enzyme production is for detergents. The use of biological enzymes in detergents is preferred as they are biodegradable, do not hamper the sewage degradation process and cause no adverse effects to aquatic life.[55] Subtilisin, an alkaline serine protease produced by *Bacillus* sp., is the predominant enzyme used in detergents. Since detergents in households are used at low temperatures, saving cost and energy, the cold-active subtilisin achieved from microorganisms could prove more economical and practical.[39] Apart from this, many other enzymes, such as proteases, amylases, lipases and cellulases, have been suggested for washing at low temperatures (20 °C).[62] Cellulases from psychrophilic bacteria can be used in the preparation of washing powders for the prevention of redeposition, cleaning purposes and colour care. Mannanases are used for removing mannan gum stains, which are otherwise very difficult to clean. Relatively thermolabile psychrophilic enzymes can be used for cleaning instruments in the food industries, as their over-activity may affect the quality of the final product. Moreover, they can be inactivated simply by using hot water.[43]

10.4.4 Molecular Biology, Genetic Engineering and Academia

To date, many alkaline phosphatase proteins have been extracted from psychrophilic bacteria such as *Arthrobacter*,[63] *Vibrio* sp., etc.[64] Alkaline phosphatase is an essential tool used in molecular biology. It removes 5'-phosphate from biotechnological vectors like bacteriophages or plasmids, preventing self-ligation while facilitating the ligation of desired products. It is also used to remove 5'-phosphate from DNA fragments before the insertion of radiolabelled phosphate. Further, the heat-labile enzymes provide the additional advantage of easy inactivation at higher temperatures but below the *Tm* of DNA. Similarly, DNA ligase is another important player in molecular biology. They are found to play an important part in DNA replication, recombination and repair mechanisms by forming a phosphodiester bond between adjacent 5′P and 3′OH. Georgette and coworkers stated that DNA ligase isolated from *Pseudoalteromonas haloplanktis* could be an important tool in biotechnology.[65] Other applications of psychrophilic enzymes include uracil-DNA N-glycosylases, which are used for the removal of uracil specifically from DNA and nucleases. The nucleases are also used for the prevention of contamination by removal of unwanted DNA from polymerase chain reaction (PCR) and RNA preparations.[66] Uracil-DNA N-glycosylase from a marine psychrophilic bacterium, *Vibrio salmonicida*, has been studied recently.[67] A novel nuclease from a psychrophilic bacterium, *Psychromonas ingrahamii* 37, has been reported to work efficiently at −12 °C.[66]

The co-expression of cold-active chaperons in *Escherichia coli* is an emerging challenge for the scientific community. However, efforts are continuously being made to elucidate the mechanism and ease the co-expression of these chaperons in *E. coli*. In one study, it was shown that the expression of cold-active chaperon DnaK from an Antarctic bacterium, *Shewanella* sp. Ac10, in *E. coli* supported its growth at low temperatures.[68] In another study, it was suggested that the homologous chaperons of *E. coli* acted as growth-limiting factors at low temperatures. The introduction of genes of *Oleispira antartica* chaperonins Cpn60 and Cpn10 in *E. coli* caused much faster growth at low temperatures.[69]

10.4.5 Pharmaceutical and Biomedical Applications

Psychrophilic lipases are important biocatalysts in biomedical reactions.[55] The preparation of single-isomer chiral drugs by the use of lipase-like enzymes in organic solvents has proved to be a promising method.[70] A reaction between stereospecific N-arylamines and lipases in *Pseudomonas* sp. forms amines, which are optically active and act as intermediates for pharmaceuticals. Applications of psychrophilic bacteria are found in the field of cosmetics also. Antarticine NF3, a hexapolymeric glycoprotein isolated from *P. antartica*, is described by Margesin and Fellar for its pharmaceutical, veterinary and cosmetics applications. This protein is an important component of many wound-healing creams.[39]

There is a plethora of knowledge available about the production of antibiotics by psychrophilic bacteria using secondary metabolism, which ensures protection to the host organism by inhibiting the growth of pathogens. The psychrophilic actinomycetes isolated from Rothang hill soil have shown antimicrobial activities against dental caries and periodontal disease-causing *Streptococcus oralis and Streptococcus mutans*.[71] Recently, Jawani and coworkers isolated four psychrophilic bacteria strains, *Psychrobacter* sp., *Aeromonas salmonicida*, *Pseudomonas veronii* and *Yersinia kristensenii*, from Antarctic waters which showed excellent antimicrobial properties by the production of silver nanoparticles.[72]

Several bacteria have a set of a minimum of a hundred genes called *essential genes*, which are pivotal for bacterial survival. Amino acid sequences of the products formed by these genes are highly conserved in all bacteria, including extremophiles such as psychrophiles. For controlling the populations of harmful bacteria, efforts have been made to delete the essential genes, which is only possible when a complementary set of genes is present. Essential genes of the bacterium *Francisella tularensis*, the causative agent of tularaemia, when substituted with their psychrophilic homologs, were able to grow at cooler parts of the body but failed to grow in internal vital organs that have higher temperatures. They were able to induce protective immunity against the disease-causing *Salmonella enterica*. When the essential genes of these mesophilic pathogens were substituted with psychrophilic homologs, the pathogen lost its pathogenicity. Therefore, this property was proved to give protective immunity against pathogens.[73]

10.4.6 Bioremediation and biodegradation

Psychrophilic microorganisms are reasonable natural tools for the bioremediation process in soils of cold regions. The pollutants are initially converted to less hazardous substances, which are further hydrolysed through the biogeochemical cycle.[39] At low temperatures, bacteria such as *Polaromonas* are responsible for the degradation of xenobiotics such as chlorinated ethenes.[74] *Alcanivorax borkumensis*, a hydrocarbon-degrading bacterium, has been described from the marine environment.[75] Polyphenols of plant origin are reported to be degraded by some psychrophilic bacteria isolated from pristine soils of Siberia.[76] Degradation of organic waste is an important property shown by psychrophilic bacteria. Bacteria isolated from the Antarctic sea showed excellent ability to degrade fats at very low temperatures.[77] Exopolysaccharide of *Pseudoalteromonas* sp. SM9913, a bacterium isolated from deep-sea sediments, can be used for flocculation in wastewater treatment. Recently, a consortium of 12 different psychrophilic bacteria has been formulated for composting.[78] These bacteria showed synergy with each other, thereby improving the rate of composting.

10.4.7 Usage in agricultural and food storage

Microorganisms have important implications for agroecosystems performing various functions such as protection from pathogens, predators, nitrogen fixation, nutrient solubilisation or mobilisation, promotion of growth, etc. Bacteria that promote plant growth are called plant growth-promoting (PGP) bacteria.[79] Various PGP bacteria that cause siderophore excretion, inhibition of phytopathogens, solubilisation of tricalcium phosphate, aluminium and iron phosphate, and production of phytohormones or HCN have been isolated from a cold habitat, the Venezuelan Andes. Inheritance of these features makes the bacteria vital biofertilisers under cold conditions.[80] Different psychrophilic bacteria screened from soil and water samples from the Leh, Ladakh region of India by Yadav and coworkers also showed plant growth promotion activities.[81] Phosphate solubilisation by microorganisms is an important aspect of plant growth. Five psychrophilic bacteria isolated from the Indian Himalayas have been tested as positive for phosphate solubilisation by Trivedi and Pandey.[82] Similarly, two psychrophilic bacteria with phosphate solubilising capability from soils of Garhwal Himalayas have been described as improving rice growth.[3] In another study, the authors stated that bacteria from genera such as *Arthrobacter*, *Flavimonas*, *Flavobacterium*, *Massilia*, *Pedobacter* and *Pseudomonas* significantly improve seed germination and growth of plants under chilling stress.[84] The researchers further added that the tested psychrophiles reduced cell membrane-associated damage and promoted antioxidant activity and proline synthesis in tomato plants under cold stress conditions.

Antifreeze protein production is one of the major strategies used by psychrophilic bacteria for growth at low temperatures. Antifreeze proteins hinder the formation of large ice crystals.[85] Though the complete industrial potential of antifreeze proteins is not well known, some probable applications could be in cryopreservation of cells, as they have the potential to interact with ice crystals and cell surface components.[86] Apart from preservation of foods, Type III antifreeze proteins have applications in CO_2 hydrate slurry production.[87] During freeze-thaw cycles, the crispness of vegetables like cucumbers and courgettes is lost; hence, antifreeze proteins may be potential candidates to preserve the taste and freshness of the vegetables.[88] They are also proved to be significant adjuvants in cryosurgery.

10.4.8 Ice formation

Ice nucleation proteins (INPs) allow nucleation of ice at relatively high temperatures, above −5 °C. In a study by Wu and colleagues,[89] a novel INP was reported from *Pseudomonas borealis* DL7, isolated from soils of tundra ecosystems of Canada. Another INP from *Pseudomonas syringae* S203 has been reported by Wolber and colleagues during the onset of research on psychrophiles.[90] As these bacteria are responsible for ice nucleation at relatively very high temperatures, they have potential application in artificial snow formation. In addition, to be inactive at high temperatures, the associated ease and sensitivity of ice nuclei make these bacteria a promising tool to study the transcription process. Further, they can be used for the biotechnological engineering of frost-resistant plant species.

10.5 Diversity and bioprospection of cold-living bacteria

As mentioned earlier, most of the surface of the Earth is covered by cold regions. The cold, harsh environment shows a predominant number of microorganisms. In permafrost soils and sea ice, it is seen that the bacteria are confined to unfrozen waters with high quantities of salts, exopolymeric substances, particulate matter and fluid flow.[91] In samples collected from soil, mud, sediment, water and ice from the Antarctic region, *Pseudomonas* and *Flavobacterium* were the dominant culturable psychrophiles.[92] Psychrophiles can also be present in refrigerated conditions. Since alpine soils show continuous biochemical and temperature shifts, they may be potential reservoirs of psychrophiles.[93] More interestingly, scientists propose that the presence of psychrophilic bacteria on the surface of Mars is possible due to the low temperature of the surroundings. Though the temperature of Mars is about −80 °C, Gilbert, the main investigator of the Viking mission to Mars in 1976, stated that microbial life does exist on Mars, which was further supported by recent studies. However, the polar microorganisms are analogues of microbial life on Mars.[94] Recently, the data curve also suggested the presence of microbial respiration on Mars. The results were similar to the results of the labelled release (LR) test conducted on soil on Earth.[95]

During the last century, many unique taxa have been reported from sea ice. Phylogenetic diversity from different genera, *Achromobacteria, Alcaligenes, Alteromonas, Aquaspirillum, Arthrobacter, Bacillus, Bacteroides, Brevibacterium, Clostridium, Colwellia, Cytophaga, Flavobacterium, Gelidibacter, Methanococcoides, Methanogenium, Methanosarcina, Microbacterium, Micrococcus, Moritella, Octadecabacter, Phormidium, Photobacterium, Polaribacter, Polaromonas, Pseudomonas, Psychroserpens, Shewanella, Psychrobacter* and *Vibrio*, has been described as psychrophilic.[96]

Principally, bioprospection of psychrophilic organisms is based on three broad approaches: culture-dependent techniques, recombinant proteins expressed in a suitable host, and metagenomic analysis. Though marine bacteria are very abundant, with a density of 10^6 ml^{-1}, the oceanic snow bacteria have a density of 10^9 ml^{-1}. However, Whitman and his group observed a tremendous number of bacteria (3.67×10^{30} ml^{-1}) associated with the marine subsurface.[97] These marine bacteria are mostly dominated by psychrophilic bacteria due to the prevailing temperature of their niche.[98] The huge yet under-explored diversity of psychrophilic bacteria has intrigued researchers to find out their ecological roles and potential applications for the production of novel enzymes and useful metabolites, though very few psychrophilic bacteria are culturable, and only 0.1% of marine bacterial diversity is culturable by conventional means.[99] However, recent advances in molecular biology tools have paved the way for deciphering the vast diversity of these cryptic microbes by using omics approaches. Metagenomics is one such approach that has emerged as an efficient tool to exploit the non-culturable microbial diversity present in marine environments.[100] This technique has allowed the microbial diversity in very minute samples and cryptic niches to be explored more successfully. Thus, this approach is suggested as advanced and useful for quantitative microbial studies in cold environments.[101] Currently, more than 315 metagenomes of psychrophiles sampled from marine or soil sediments are available in the MG-RAST database.[2] The diversity and community structure of psychrophilic bacteria from glaciers of the German Alps region has been studied.[102] In the metagenomic analysis of Antarctic soil, carboxymethyl cellulose-degrading *Pseudomonas stutzeri* was reported.[103] Similarly, metagenomics of tidal flat sediment resulted in the evaluation of a novel psychrophilic alkaline phosphatase.[104]

10.6 CONCLUSION

Psychrophilic bacteria form a significant part of the biosphere, showing presence even in extreme cold places such as the poles, mountains and deep oceans. The lowest temperature limit reported for bacteria is –20 °C. Apart from psychrophilic bacteria, cold-tolerant and cold-adapted enzymes are also important from a biotechnological standpoint. They offer numerous advantages in industries like food, laundry, etc. Additionally, psychrophilic enzymes have significance in molecular biology, genetic engineering and other pharmaceutical industries. Bioremediation of pollutants, agricultural usage and food storage are other processes where psychrophilic bacteria are important. Due to their potential applications, continuous bioprospection of efficient psychrophilic bacteria and their enzymes is an important area of research. Therefore, there is a constant impetus to search for novel psychrophilic bacteria with advanced techniques like metagenomics and metaproteomics. In addition, the engineering of the cold-adapted enzymes secreted by some of the psychrophilic bacteria is greatly needed to meet the challenges of the future food industry.

ACKNOWLEDGEMENTS

The authors are thankful to the Council of Scientific and Industrial Research (CSIR), New Delhi, for fellowship. The funding provided by National Natural Science Foundation of China (32250410285) and Priority of Academic Program Development of Jiangsu Higher Education Institutions (PAPD 4013000011) is also acknowledged.

REFERENCES

1. Barnes, David K.A., Francois Galgani, Richard C. Thompson, and Morton Barlaz. "Accumulation and fragmentation of plastic debris in global environments." *Philosophical Transactions of the Royal Society B: Biological Sciences* 364, no. 1526 (2009): 1985–1998.
2. De Maayer, Pieter, Dominique Anderson, Craig Cary, and Don A. Cowan. "Some like it cold: Understanding the survival strategies of psychrophiles." *EMBO reports* 15, no. 5 (2014): 508–517.
3. Collins, Tony, and Rosa Margesin. "Psychrophilic lifestyles: Mechanisms of adaptation and biotechnological tools." *Applied Microbiology and Biotechnology* 103, no. 7 (2019): 2857–2871.
4. Price, P. Buford, and Todd Sowers. "Temperature dependence of metabolic rates for microbial growth, maintenance, and survival." *Proceedings of the National Academy of Sciences* 101, no. 13 (2004): 4631–4636.
5. Clarke, Andrew, G. John Morris, Fernanda Fonseca et al. "A low temperature limit for life on Earth." *PLoS One* 8, no. 6 (2013): e66207.
6. Siddiqui, Khawar Sohail, and Ricardo Cavicchioli. "Cold-adapted enzymes." *Annual Review of Biochemistry* 75 (2006): 403–433.
7. Kottmeier, Steven T., and Cornelius W. Sullivan. "Bacterial biomass and production in pack ice of Antarctic marginal ice edge zones." *Deep Sea Research Part A: Oceanographic Research Papers* 37, no. 8 (1990): 1311–1330.
8. Chaudhary, Dhiraj Kumar, Dong-Uk Kim, Dockyu Kim, and Jaisoo Kim. "*Flavobacterium petrolei* sp. nov., a novel psychrophilic, diesel-degrading bacterium isolated from oil-contaminated Arctic soil." *Scientific Reports* 9, no. 1 (2019): 1–9.
9. Godoy, F., Marc Vancanneyt, M. Martinez et al. "*Sphingopyxischilensis* sp. nov., a chlorophenol-degrading bacterium that accumulates polyhydroxyalkanoate, and transfer of Sphingomonas alaskensis to Sphingopyxis alaskensis comb. nov."

International Journal of Systematic and Evolutionary Microbiology 53, no. 2 (2003): 473–477.

10. Butler, Barbara J., Kirk L. Mccallum, and William E. Inniss. "Characterization of *Aquaspirillum arcticum* sp. nov., a new psychrophilic bacterium." *Systematic and Applied Microbiology* 12, no. 3 (1989): 263–266.
11. Reddy, G. S., R. K. Aggarwal, G. I. Matsumoto, and S. Shivaji. "*Arthrobacter flavus* sp. nov., a psychrophilic bacterium isolated from a pond in McMurdo Dry Valley, Antarctica." *International Journal of Systematic and Evolutionary Microbiology* 50, no. 4 (2000): 1553–1561.
12. Margesin, Rosa, Peter Schumann, De-Chao Zhang et al. "*Arthrobacter cryoconiti* sp. nov., a psychrophilic bacterium isolated from alpine glacier cryoconite." *International Journal of Systematic and Evolutionary Microbiology* 62, no. 2 (2012): 397–402.
13. Margesin, Rosa, Peter Schumann, Cathrin Spröer, and Anne-Monique Gounot. "*Arthrobacter psychrophenolicus* sp. nov., isolated from an alpine ice cave." *International Journal of Systematic and Evolutionary Microbiology* 54, no. 6 (2004): 2067–2072
14. Rüger, H. J., D. Fritze, and C. Spröer. "New psychrophilic and psychrotolerant *Bacillus marinus* strains from tropical and polar deep-sea sediments and emended description of the species." *International Journal of Systematic and Evolutionary Microbiology* 50, no. 3 (2000): 1305–1313.
15. Zhang, Cong, Wenbin Guo, Yuguang Wang, and Xinhua Chen. "*Colwellia beringensis* sp. nov., a psychrophilic bacterium isolated from the Bering Sea." *International Journal of Systematic and Evolutionary Microbiology* 67, no. 12 (2017): 5102–5107.
16. Gosink, J. J., R. P. Herwig, and J. T. Staley. "*Octadecabacter arcticus* gen. nov., sp. nov., and *O. antarcticus* sp. nov., nonpigmented, psychrophilic gas vacuolate bacteria from polar sea ice and water." *Systematic and Applied Microbiology* 20, no. 3 (1997): 356–365.
17. Morita, Richard Y. "Psychrophilic bacteria." *Bacteriological Reviews* 39, no. 2 (1975): 144–167.
18. Mitra, Debanjan, and Pradeep Kr Das Mohapatra. "Cold adaptation strategy of psychrophilic bacteria: An in-silico analysis of isocitrate dehydrogenase." *Systems Microbiology and Biomanufacturing* 1, no. 4 (2021): 483–493.
19. Rivkina, E. M., E. I. Friedmann, C. P. McKay, and D. A. Gilichinsky. "Metabolic activity of permafrost bacteria below the freezing point." *Applied and Environmental Microbiology* 66, no. 8 (2000): 3230–3233.
20. Youle, Merry, Forest Rohwer, Apollo Stacy et al. "The microbial olympics." *Nature Reviews Microbiology* 10, no. 8 (2012): 583–588.
21. Keenleyside, W. "Microbiology: Canadian edition." Open Library Press Books. Available online: https://ecampusontario.pressbooks.pub/microbio/front-matter/preface (accessed on 20 August 2020) (2019).
22. Phadtare, Sangita. "Recent developments in bacterial cold-shock response." *Current Issues in Molecular Biology* 6, no. 2 (2004): 125–136.
23. Membré, Jeanne-Marie, Benoît Leporq, Michele Vialette et al. "Temperature effect on bacterial growth rate: Quantitative microbiology approach including cardinal values and variability estimates to perform growth simulations on/in food." *International Journal of Food Microbiology* 100, no. 1–3 (2005): 179–186.
24. Whiting, R. C., and L. K. Bagi. "Modeling the lag phase of Listeria monocytogenes." *International Journal of Food Microbiology* 73, no. 2–3 (2002): 291–295.

25. Mansilla, María C., Larisa E. Cybulski, Daniela Albanesi, and Diego de Mendoza. "Control of membrane lipid fluidity by molecular thermosensors." *Journal of Bacteriology* 186, no. 20 (2004): 6681–6688.
26. Koeberle, Andreas, Hideo Shindou, Takeshi Harayama, and Takao Shimizu. "Palmitoleate is a mitogen, formed upon stimulation with growth factors, and converted to palmitoleoyl-phosphatidylinositol." *Journal of Biological Chemistry* 287, no. 32 (2012): 27244–27254.
27. Cao-Hoang, L., Frédéric Dumont, Pierre-André Marechal, and Patrick Gervais. "Inactivation of *Escherichia coli* and *Lactobacillus plantarum* in relation to membrane permeabilization due to rapid chilling followed by cold storage." *Archives of Microbiology* 192, no. 4 (2010): 299–305.
28. Barria, C., M. Malecki, and C. M. Arraiano. "Bacterial adaptation to cold." *Microbiology* 159, no. Pt_12 (2013): 2437–2443.
29. Di Pietro, Fabio, Anna Brandi, Nadire Dzeladini et al. "Role of the ribosome-associated protein PY in the cold-shock response of *Escherichia coli*." *Microbiology Open* 2, no. 2 (2013): 293–307.
30. Russell, N. J. *Advances in Biochemical Engineering/Biotechnology* (1998): 1–21.
31. Inniss, William E. "Interaction of temperature and psychrophilic microorganisms." *Annual Review of Microbiology* 29, no. 1 (1975): 445–466.
32. Jin, Shanshan, Yizhe Wang, and Xihong Zhao. "Cold-adaptive mechanism of psychrophilic bacteria in food and its application." *Microbial Pathogenesis* (2022): 105652.
33. Králová, Stanislava. "Role of fatty acids in cold adaptation of Antarctic psychrophilic *Flavobacterium* spp." *Systematic and Applied Microbiology* 40, no. 6 (2017): 329–333.
34. Borchert, Erik, Stephen Knobloch, Emilie Dwyer et al. "Biotechnological potential of cold adapted *Pseudoalteromonas* spp. isolated from "deep sea"sponges." *Marine Drugs* 15, no. 6 (2017): 184.
35. Abraham, Wilson Peter, Sajith Raghunandanan, Vipin Gopinath et al. "Deciphering the cold adaptive mechanisms in *Pseudomonas psychrophila* MTCC12324 isolated from the Arctic at 79 N." *Current Microbiology* 77, no. 9 (2020): 2345–2355.
36. Humphry, David R., Alison George, Gary W. Black, and Stephen P. Cummings. "*Flavobacterium frigidarium* sp. nov., an aerobic, psychrophilic, xylanolytic and laminarinolytic bacterium from Antarctica." *International Journal of Systematic and Evolutionary Microbiology* 51, no. 4 (2001): 1235–1243.
37. Mishra, Pankaj Kumar, Piyush Joshi, Shekhar Chandra Bisht et al. "Cold-tolerant agriculturally important microorganisms." In *Plant Growth and Health Promoting Bacteria*, pp. 273–296. Springer, Berlin, Heidelberg, (2010).
38. Liu, Siqing, James E. Graham, Lance Bigelow et al. "Identification of Listeria monocytogenes genes expressed in response to growth at low temperature." *Applied and Environmental Microbiology* 68, no. 4 (2002): 1697–1705.
39. Margesin, Rosa, and Georges Feller. "Biotechnological applications of psychrophiles." *Environmental Technology* 31, no. 8–9 (2010): 835–844.
40. Shivaji, S., and Jogadhenu SS Prakash. "How do bacteria sense and respond to low temperature?." *Archives of Microbiology* 192, no. 2 (2010): 85–95.
41. Dieser, Markus, Mark Greenwood, and Christine M. Foreman. "Carotenoid pigmentation in Antarctic heterotrophic bacteria as a strategy to withstand environmental stresses." *Arctic, Antarctic, and Alpine Research* 42, no. 4 (2010): 396–405.
42. McGann, Patrick, Renata Ivanek, Martin Wiedmann, and Kathryn J. Boor. "Temperature-dependent expression of Listeria monocytogenes internalin and

internalin-like genes suggests functional diversity of these proteins among the listeriae." *Applied and Environmental Microbiology* 73, no. 9 (2007): 2806–2814.
43. Cavicchioli, R., T. Charlton, H. Ertan et al. "Biotechnological uses of enzymes from psychrophiles." *Microbial Biotechnology* 4, no. 4 (2011): 449–460.
44. Yadav, Ajar Nath, Shashwati Ghosh Sachan, Priyanka Verma, and Anil Kumar Saxena. "Bioprospecting of plant growth promoting psychrotrophic *Bacilli* from the cold desert of north western Indian Himalayas." 54 (2016): 142–150.
45. Kim, Hyun, Ae Kyung Park, Jun Hyuck Lee et al. "PsEst3, a new psychrophilic esterase from the Arctic bacterium *Paenibacillus* sp. R4: Crystallization and X-ray crystallographic analysis." *Acta Crystallographica Section F: Structural Biology Communications* 74, no. 6 (2018): 367–372.
46. De Santi, Concetta, Pietro Tedesco, Luca Ambrosino et al. "A new alkaliphilic cold-active esterase from the psychrophilic marine bacterium *Rhodococcus* sp.: Functional and structural studies and biotechnological potential." *Applied Biochemistry and Biotechnology* 172, no. 6 (2014): 3054–3068.
47. Salwoom, Leelatulasi, Raja Noor Zaliha Raja Abd Rahman, Abu Bakar Salleh et al. "Isolation, characterisation, and lipase production of a cold-adapted bacterial strain *Pseudomonas* sp. LSK25 isolated from Signy Island, Antarctica." *Molecules* 24, no. 4 (2019): 715.
48. Wang, Fengping, Yingbao Gai, Mingxia Chen, and Xiang Xiao. "*Arthrobacter psychrochitiniphilus* sp. nov., a psychrotrophic bacterium isolated from Antarctica." *International Journal of Systematic and Evolutionary Microbiology* 59, no. 11 (2009): 2759–2762
49. Zeng, Runying, Pengjun Xiong, and Jianjun Wen. "Characterization and gene cloning of a cold-active cellulase from a deep-sea psychrotrophic bacterium *Pseudoalteromonas* sp. DY3." *Extremophiles* 10, no. 1 (2006): 79–82.
50. Shipkowski, S., and Brenchley, J. E. (2005). Characterization of an unusual cold-active β-glucosidase belonging to family 3 of the glycoside hydrolases from the psychrophilic isolate *Paenibacillus* sp. strain C7. *Applied and Environmental Microbiology*, *71*(8), 4225–4232.
51. Van Truong, Le, Ho Tuyen, Elisabeth Helmke et al. "Cloning of two pectate lyase genes from the marine Antarctic bacterium *Pseudoalteromonas haloplanktis* strain ANT/505 and characterization of the enzymes." *Extremophiles* 5, no. 1 (2001): 35–44.
52. Orikoshi, Hideyuki, Nao Baba, Shigenari Nakayama, Hiroshi Kashu, Katsushiro Miyamoto, Masahide Yasuda, Yoshihiko Inamori, and Hiroshi Tsujibo. "Molecular analysis of the gene encoding a novel cold-adapted chitinase (ChiB) from a marine bacterium, *Alteromonas* sp. strain O-7." *Journal of Bacteriology* 185, no. 4 (2003): 1153–1160.
53. Lonhienne, Thierry, Konstantinos Mavromatis, Constantin E. Vorgias et al. "Cloning, sequences, and characterization of two chitinase genes from the Antarctic *Arthrobacter* sp. strain TAD20: Isolation and partial characterization of the enzymes." *Journal of Bacteriology* 183, no. 5 (2001): 1773–1779.
54. Gerday, Charles, Mohamed Aittaleb, Mostafa Bentahir et al. "Cold-adapted enzymes: from fundamentals to biotechnology." *Trends in Biotechnology* 18, no. 3 (2000): 103–107.
55. Joseph, Babu, Pramod W. Ramteke, and George Thomas. "Cold active microbial lipases: Some hot issues and recent developments." *Biotechnology Advances* 26, no. 5 (2008): 457–470.
56. Joshi, Swati, and Tulasi Satyanarayana. "Biotechnology of cold-active proteases." *Biology* 2, no. 2 (2013): 755–783.

57. Kuddus, Mohammed, and Pramod W. Ramteke. "Recent developments in production and biotechnological applications of cold-active microbial proteases." *Critical Reviews in Microbiology* 38, no. 4 (2012): 330–338.
58. Aurilia, Vincenzo, Antonietta Parracino, and Sabato D'Auria. "Microbial carbohydrate esterases in cold adapted environments." *Gene* 410, no. 2 (2008): 234–240.
59. Pedrolli, Danielle Biscaro, Alexandre Costa Monteiro, Eleni Gomes, and Eleonora Cano Carmona. "Pectin and pectinases: Production, characterization and industrial application of microbial pectinolytic enzymes." *Open Biotechnology Journal* (2009): 9–18.
60. Adapa, Vijayanand, L. N. Ramya, K. K. Pulicherla, and K. R. S. Rao. "Cold active pectinases: Advancing the food industry to the next generation." *Applied Biochemistry and Biotechnology* 172, no. 5 (2014): 2324–2337.
61. Wang, Quan-Fu, Yan-Hua Hou, Zhong Xu et al. "Purification and properties of an extracellular cold-active protease from the psychrophilic bacterium *Pseudoalteromonas* sp. NJ276." *Biochemical Engineering Journal* 38, no. 3 (2008): 362–368.
62. Aehle, Wolfgang, ed. *Enzymes in Industry: Production and Applications.* John Wiley & Sons, 2007.
63. De Prada, Paloma, Jennifer Loveland-Curtze, and Jean E. Brenchley. "Production of two extracellular alkaline phosphatases by a psychrophilic Arthrobacter strain." *Applied and Environmental Microbiology* 62, no. 10 (1996): 3732–3738.
64. Hauksson, Jónas B., Ólafur S. Andrésson, and Bjarni Ásgeirsson. "Heat-labile bacterial alkaline phosphatase from a marine *Vibrio* sp." *Enzyme and Microbial Technology* 27, no. 1–2 (2000): 66–73.
65. Georlette, Daphné, Benjamin Damien, Vinciane Blaise et al. "Structural and functional adaptations to extreme temperatures in psychrophilic, mesophilic, and thermophilic DNA ligases." *Journal of Biological Chemistry* 278, no. 39 (2003): 37015–37023.
66. Maciejewska, Natalia, Rafał Walkusz, Marcin Olszewski, and Aneta Szymańska. "New nuclease from extremely psychrophilic microorganism *Psychromonas ingrahamii* 37: Identification and characterization." *Molecular Biotechnology* 61, no. 2 (2019): 122–133.
67. Ræder, Inger Lin Uttakleiv, Ingar Leiros et al. "Uracil-DNA N-glycosylase (UNG) from the marine, psychrophilic bacterium Vibrio salmonicida shows cold adapted features: A comparative analysis to Vibrio cholerae uracil-DNA N-glycosylase." *Enzyme and Microbial Technology* 42, no. 7 (2008): 594–600.
68. Yoshimune, Kazuaki, Andrey Galkin, Ljudmila Kulakova et al. "Cold-active DnaK of an Antarctic psychrotroph *Shewanella* sp. Ac10 supporting the growth of dnaK-null mutant of *Escherichia coli* at cold temperatures." *Extremophiles* 9, no. 2 (2005): 145–150.
69. Ferrer, Manuel, Tatyana N. Chernikova, Michail M. Yakimov et al. "Chaperonins govern growth of *Escherichia coli* at low temperatures." *Nature Biotechnology* 21, no. 11 (2003): 1266–1267.
70. Gotor-Fernández, Vicente, Rosario Brieva, and Vicente Gotor. "Lipases: Useful biocatalysts for the preparation of pharmaceuticals." *Journal of Molecular Catalysis B: Enzymatic* 40, no. 3–4 (2006): 111–120.
71. Raja, A., P. Prabakaran, and P. Gajalakshmi. "Isolation and screening of antibiotic producing psychrophilic actinomycetes and its nature from rothang hill soil against viridans *Streptococcus* sp." *Research Journal of Microbiology* 5, no. 1 (2010): 44–49.
72. Javani, Siamak, Irma Marín, Ricardo Amils, and José Pascual Abad. "Four psychrophilic bacteria from Antarctica extracellularly biosynthesize at low temperature

highly stable silver nanoparticles with outstanding antimicrobial activity." *Colloids and Surfaces A: Physicochemical and Engineering Aspects* 483 (2015): 60–69.
73. Duplantis, Barry N., Stephanie M. Puckett, Everett L. Rosey et al. "Temperature-sensitive *Salmonella enterica* serovar enteritidis PT13a expressing essential proteins of psychrophilic bacteria." *Applied and Environmental Microbiology* 81, no. 19 (2015): 6757–6766.
74. Mattes, Timothy E., Anne K. Alexander, Paul M. Richardson et al. "The genome of *Polaromonas* sp. strain JS666: Insights into the evolution of a hydrocarbon- and xenobiotic-degrading bacterium, and features of relevance to biotechnology." *Applied and Environmental Microbiology* 74, no. 20 (2008): 6405–6416.
75. Golyshin, Peter N., Vitor AP Martins Dos Santos, Olaf Kaiser et al. "Genome sequence completed of *Alcanivorax borkumensis*, a hydrocarbon-degrading bacterium that plays a global role in oil removal from marine systems." *Journal of Biotechnology* 106, no. 2–3 (2003): 215–220.
76. Koutny, Marek, Jan Ruzicka, and Jiri Chlachula. "Screening for phenol-degrading bacteria in the pristine soils of south Siberia." *Applied Soil Ecology* 23, no. 1 (2003): 79–83.
77. Feiler, K. "Trematoden aus Chionis alba und Larus dominicanus von den Süd-Shetland-Inseln (Antarktis)." *Angewandte Parasitologie* 27, no. 1 (1986): 23–33.
78. Hou, Ning, Luming Wen, Huiming Cao et al. "Role of psychrotrophic bacteria in organic domestic waste composting in cold regions of China." *Bioresource Technology* 236 (2017): 20–28.
79. Andrews, John H., and Robin F. Harris. "The ecology and biogeography of microorganisms on plant surfaces." *Annual Review of Phytopathology* 38 (2000): 145.
80. Balcazar, Wilvis, Johnma Rondón, Marcos Rengifo et al. "Bioprospecting glacial ice for plant growth promoting bacteria." *Microbiological Research* 177 (2015): 1–7.
81. Yadav, Ajar Nath, Shashwati Ghosh Sachan, Priyanka Verma et al. "Culturable diversity and functional annotation of psychrotrophic bacteria from cold desert of Leh Ladakh (India)." *World Journal of Microbiology and Biotechnology* 31, no. 1 (2015): 95–108.
82. Trivedi, Pankaj, and Anita Pandey. "Low temperature phosphate solubilization and plant growth promotion by psychrotrophic bacteria, isolated from Indian Himalayan region." *Research Journal of Microbiology* (2007): 454–461.
83. Gusain, Yogendra Singh, Ranveer Kamal, C. M. Mehta et al. "Phosphate solubilizing and indole-3-acetic acid producing bacteria from the soil of Garhwal Himalaya aimed to improve the growth of rice." *Journal of Environmental Biology* 36, no. 1 (2015): 301.
84. Subramanian, Parthiban, Kiyoon Kim, Ramasamy Krishnamoorthy et al. "Cold stress tolerance in psychrotolerant soil bacteria and their conferred chilling resistance in tomato (*Solanum lycopersicum* Mill.) under low temperatures." *PloS One* 11, no. 8 (2016): e0161592.
85. Lorv, Janet SH, David R. Rose, and Bernard R. Glick. "Bacterial ice crystal controlling proteins." *Scientifica* 2014 (2014).
86. Jeon, Su Min, Aung Htay Naing, Kyeung Il Park, and Chang Kil Kim. "The effect of antifreeze protein on the cryopreservation of Chrysanthemums." *Plant Cell, Tissue and Organ Culture (PCTOC)* 123, no. 3 (2015): 665–671.
87. Zhou, Hongxia, and Carlos A. Infante Ferreira. "The effect of type-III antifreeze proteins (AFPs) on CO2 hydrate slurry formation." (2014).
88. Muñoz, Patricio A., Sebastián L. Márquez, Fernando D. González-Nilo, Valeria Márquez-Miranda, and Jenny M. Blamey. "Structure and application of antifreeze proteins from Antarctic bacteria." *Microbial Cell Factories* 16, no. 1 (2017): 1–13.

89. Wu, Zhongqin, Lei Qin, and Virginia K. Walker. "Characterization and recombinant expression of a divergent ice nucleation protein from '*Pseudomonas borealis*'." *Microbiology* 155, no. 4 (2009): 1164–1169.
90. Wolber, Paul K., Caroline A. Deininger, Maurice W. Southworth et al. "Identification and purification of a bacterial ice-nucleation protein." *Proceedings of the National Academy of Sciences* 83, no. 19 (1986): 7256–7260.
91. D'Amico, Salvino, Tony Collins, Jean-Claude Marx et al. "Psychrophilic microorganisms: Challenges for life." *EMBO Reports* 7, no. 4 (2006): 385–389.
92. González-Rocha, Gerardo, Gabriel Muñoz-Cartes, Cristian B. Canales-Aguirre et al. "Diversity structure of culturable bacteria isolated from the Fildes Peninsula (King George Island, Antarctica): A phylogenetic analysis perspective." *PLoS One* 12, no. 6 (2017): e0179390.
93. Greenland, David, and Mark Losleben. "*Climate*". In Bowman R.D., Seestedt T.R. (Eds) *Structure and Function of an Alpine Ecosystem*, pp. 15–31. Niwot Ridge: Oxford University Press, CO., 2001.
94. Levin, Gilbert V. "I'm convinced we found evidence of life on mars in the 1970s." *Scientific American* 10 (2019).
95. Pikuta, Elena V., Richard B. Hoover, and Jane Tang. "Microbial extremophiles at the limits of life." *Critical Reviews in Microbiology* 33, no. 3 (2007): 183–209.
96. Hamdan, A. (2018). Psychrophiles: Ecological significance and potential industrial application. *South African Journal of Science*, *114*(5–6), 1–6.
97. Whitman, William B., David C. Coleman, and William J. Wiebe. "Prokaryotes: The unseen majority." *Proceedings of the National Academy of Sciences* 95, no. 12 (1998): 6578–6583.
98. Connelly, T. L., C. M. Tilburg, and P. L. Yager. "Evidence for psychrophiles outnumbering psychrotolerant marine bacteria in the springtime coastal Arctic." *Limnology and Oceanography* 51, no. 2 (2006): 1205–1210.
99. Kennedy, Jonathan, Julian R. Marchesi, and Alan DW Dobson. "Marine metagenomics: Strategies for the discovery of novel enzymes with biotechnological applications from marine environments." *Microbial Cell Factories* 7, no. 1 (2008): 1–8.
100. Cowan, Don, Quinton Meyer, William Stafford, Samson Muyanga, Rory Cameron, and Pia Wittwer. "Metagenomic gene discovery: past, present and future." *Trends in Biotechnology* 23, no. 6 (2005): 321–329.
101. Von Mering, C., P. Hugenholtz, Jeroen Raes et al. "Quantitative phylogenetic assessment of microbial communities in diverse environments." *Science* 315, no. 5815 (2007): 1126–1130.
102. Simon, Carola, Arnim Wiezer, Axel W. Strittmatter, and Rolf Daniel. "Phylogenetic diversity and metabolic potential revealed in a glacier ice metagenome." *Applied and Environmental Microbiology* 75, no. 23 (2009): 7519–7526.
103. Berlemont, Renaud, Maud Delsaute, Delphine Pipers et al. "Insights into bacterial cellulose biosynthesis by functional metagenomics on Antarctic soil samples." *The ISME Journal* 3, no. 9 (2009): 1070–1081.
104. El-Shishtawy, Reda M., Saleh A. Mohamed et al. "Saccharification and hydrolytic enzyme production of alkali pre-treated wheat bran by *Trichoderma virens* under solid state fermentation." *BMC Biotechnology* 15, no. 1 (2015): 1–13.

11 Psychrophilic soil microorganisms

A potential source of secondary metabolites

Muhammad Hayat[#], Muhammad Usman[#], Muhammad Rafiq, Imran Khan, Wasim Sajjad, Muhammad Ibrar, Aqib Sayyed and Fariha Hasan**

CONTENTS

* Corresponding authors e-mail: mr14311@my.bristol.ac.uk; farihahasan@yahoo.com
# Authors have equal contributions

DOI: 10.1201/9781003354031-11

11.1 INTRODUCTION TO PSYCHROPHILES

Approximately 80% of the Earth is at a temperature <5 °C. It covers most of the world's seas, polar zones, deep-water lakes, lake ice, cloudy atmosphere, permafrost and cold soil. It also covers Antarctica, Arctic ecosystems, stratosphere and mesosphere, and mountains such as the Rockies, Alps and Himalayas.[1] These constitute the major low-temperature ecosystems, the most abundant and broadly scattered ecosystems on the planet.[2] Ice caps, cold desert land and shallow seas surround Antarctica. It is identified as the most hostile environment on earth. Of 13,661,000 km^2, only 0.36% of Antarctica is ice-free, the majority of which is permafrost, ground that remains at a cold temperature of ≤0 °C.[3] The polar soil is underexplored in habitats having microorganisms of biotechnological importance.[4] Outside the polar regions, the Qinghai–Tibet Plateau, Himalaya, Karakoram and Hindu-Kush (HKKH) are the areas most widely considered as the third pole of the Earth. The HKKH represents more than 20,000 glaciers, of which >12,000 lie in the Himalayas region, which covers almost 60,000 km^2,[5] while in Karakoram, about 5000 glaciers are situated in this range.[6] There is a scarcity of data regarding the HKKH glaciers' microbial diversity and their biotechnological importance.

Microorganisms that inhabit low-temperature habitats are commonly known as psychrophiles or psychrotrophs. Psychrophiles and psychrotrophs were first described as microorganisms that can grow at or near 0 °C. These organisms are further classified as *sensu stricto*, which grow best at 15 °C[7] and prevail in the marine ecosystem, where the temperature remains cold at 5 °C permanently. The terrestrial environment's microorganisms are highly vulnerable to extreme environmental variations and are psychrotolerant.[8,9] However, these organisms grow at a temperature that ranges from 0 to 25 °C. The optimum temperature for maximum growth is about 25 °C.[7] There is no clear temperature limit for psychrophiles; however, a minimum of −20 °C for metabolic function and −12 °C for active division has been suggested.[10]

11.2 STRATEGIES FOR SURVIVAL

Cold-temperature microorganisms face diverse stress conditions. This environment negatively regulates enzyme kinetics, macromolecular interaction, membrane fluidity, solute diffusion rates and water viscosity,[2] maintaining vital cellular function and survival in such harsh habitats.

The microbes have overcome the stress conditions, including desiccation, high osmotic pressure, irradiation, extreme pH and poor nutritional supply, and have developed many adaptive mechanisms.[11–14] Antibiotic development is another necessary adaptation in the cold climate, allowing microorganisms to minimise interspecies competition in such a restricted nutritional supply for their entire life period.[15] The adaptation mechanisms in cold microorganisms include the production of compatible solutes, cold shock and antifreeze proteins, exopolysaccharides and polyunsaturated fatty acids,[10, 16, 17] as illustrated in Figures 11.1 and 11.2.

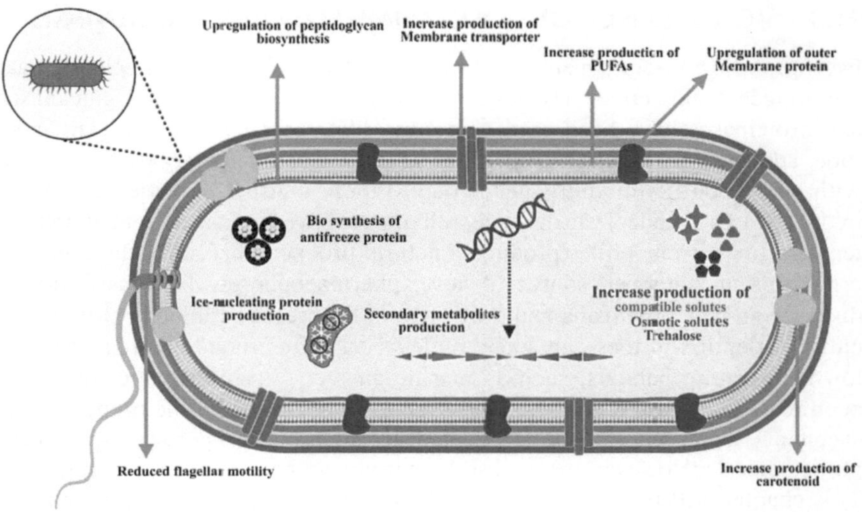

FIGURE 11.1 Adaptation mechanisms in bacteria (psychrophilic) residing in a cold environment.

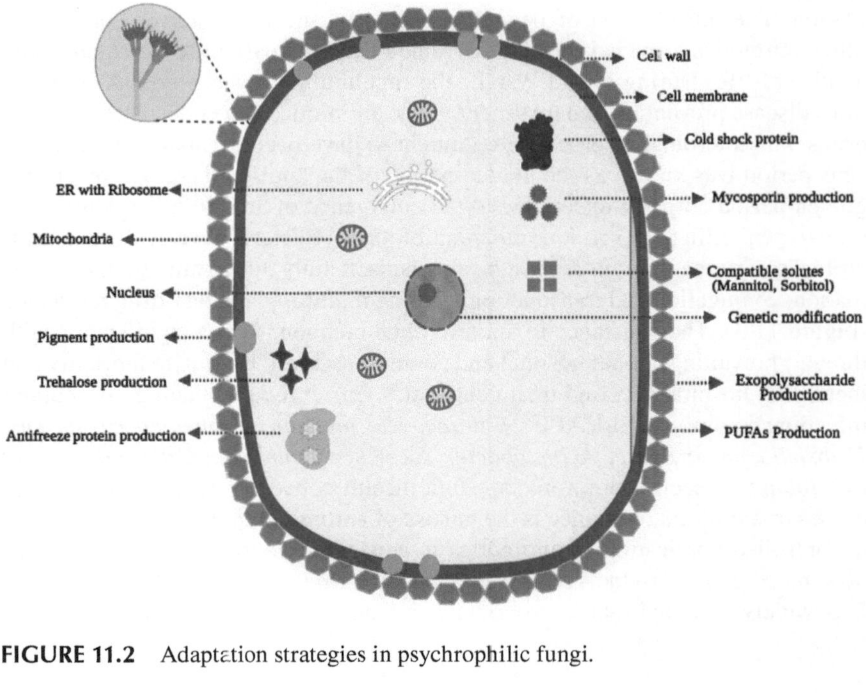

FIGURE 11.2 Adaptation strategies in psychrophilic fungi.

11.3 SIGNIFICANCE OF COLD-ADAPTED MICROORGANISMS

Psychrophilic microorganisms produce cold-active enzymes of biotechnological importance. Cold-active enzymes comprising lipases, pectinases, β-glucanases and proteinases have been used to produce detergent on an industrial scale, food additives and biosensors.[18] Commercial manufacturers are concerned with psychrophilic microorganisms due to these unusual activities. However, in the last two decades, various research reports have been documented on low-temperature extremophiles producing natural products that make these micro-organisms an untapped source of novel pharmacophores. The possibility of discovering new organisms and the potential for useful antibiotics prompt scientists' attention to these untapped outlets.[4,19] Antimicrobial compounds from low-temperature habitats, such as marine and terrestrial environments, have been recorded in a few studies. These reports [19–21] presented the production of secondary metabolites showing efficient antimicrobial properties by psychrophilic bacteria. However, the cold soil environment has not been well explored. This chapter will thoroughly discuss the bioactive metabolites from psychrophiles residing in cold soil.

11.4 ANTIBIOTIC REQUIREMENTS?

During the pre-antibiotic era, infectious diseases were a significant health issue. Numerous infectious diseases, including mumps, smallpox, cholera, syphilis, tuberculosis and plague, were the main reasons for mortality in the human population. The introduction of penicillin in 1940[21] and the discovery of various other antibiotics provided an effective remedy and increased life expectancy and quality of life. During World War II, the first antibiotic was essential in infectious disease prevention and treatment.[22] The antibiotics were considered miracle drugs, and microbial diseases were thought to have been permanently defeated. This period was known as the golden period of the antibiotic era. However, this golden period abruptly ended due to the emergence of antibiotic resistance, not only to penicillin but to various other antibiotics.[23] Bacteria have evolved antimicrobial resistance through diverse mechanisms, mainly due to mutation, transformation, conjugation and transduction, gaining multidrug and pan-drug resistance (Figure 11.3). The resistance to antimicrobial compounds is a significant health threat,[24] prevailing at both hospital and community level, leading to mortality and increasing hospital stay and treatment cost.[25] Currently, the leading nosocomial infections known as 'ESKAPE' (*Enterococcus faecium*, *Staphylococcus aureus*, *Klebsiella pneumoniae*, *Acinetobacter* species, *Pseudomonas aeruginosa* and *Enterobacter* species) are a major public health concern.[26] One of the primary causes of antibiotic resistance is the misuse of antimicrobial medications and the uncontrolled application of antibiotics in aquaculture/agriculture fields.[24] Until now, most natural products have been isolated from the mesophilic environment. It is widely assumed that microorganisms from various environments, such as

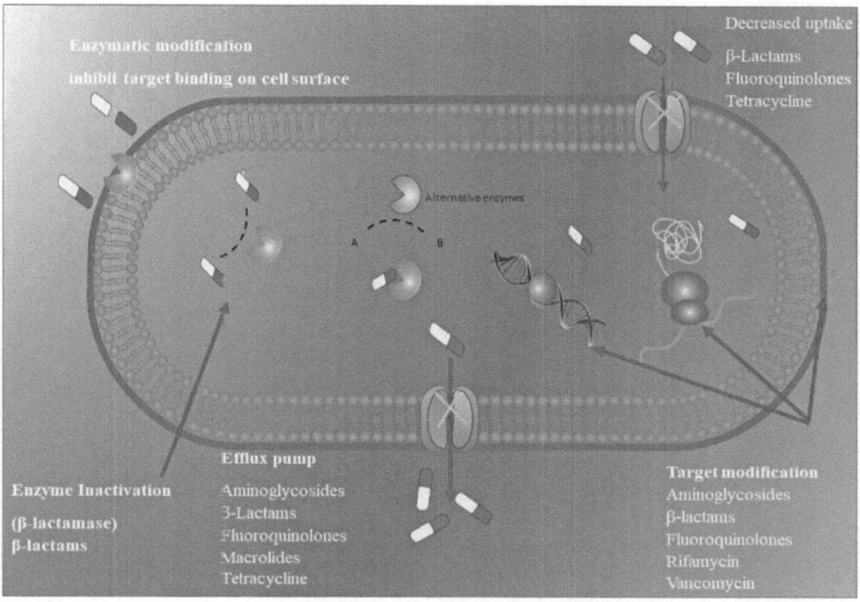

FIGURE 11.3 Molecular modifications in bacteria causing resistance to antibiotics.

those from low-temperature habitats like glaciers and soil, need to be explored and investigated to discover unusual bioactive compounds with novel modes of action

11.5 KNOWN COMPOUNDS FROM PSYCHROPHILIC SOIL FUNGI

11.5.1 Psychrophillins

Psychrophillins are a group of cyclic peptides comprising three amino acids that contain the nitro group instead of the amino group. They are composed of psychrophillins A, B, C and D. Psychrophillin A was isolated from *Penicillium reibeum* (IBT 16537), a psychrotolerant fungus isolated from the soil beneath a redcurrant bush from Wyoming Tundra, USA.[27] Psychrophillins B and C were isolated from another psychrotolerant fungus, *Penicillium rivulum,* a soil isolate maintained by the Technical University of Denmark culture collection (IBT) at Biocenturum-DTU.[27] Another cyclic neuropeptide compound, Psychrophillin D, was isolated from another psychrotolerant soil fungus, *Penicillium algidium* (IBT 22067), isolated from east of Oksestien, Zackenberg, Greenland.[28] These compounds are biologically inactive except for psychrophillin D, which is active against murine leukaemia P-388 cells.

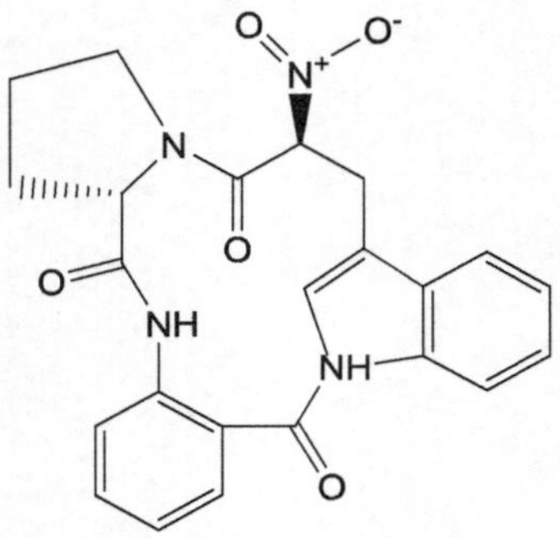

1. Psychrophilins A

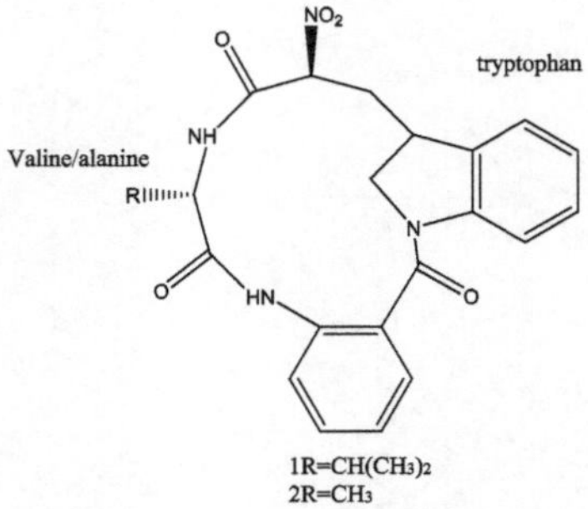

2. Psychrophilins B (1) and C (2)

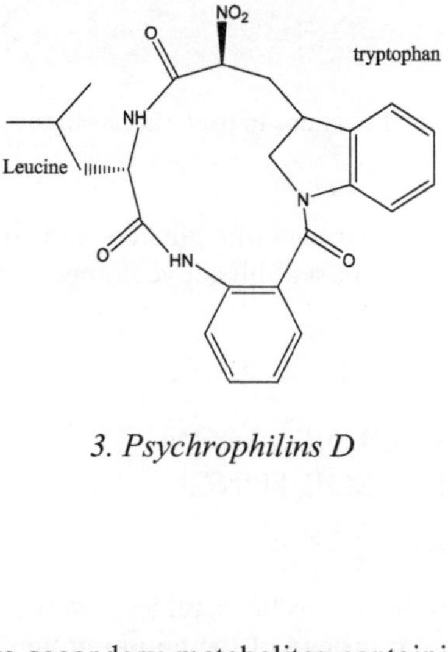

3. Psychrophilins D

11.5.2 Alkaloids

Alkaloids are bioactive secondary metabolites containing nitrogen atoms produced from common amino acids like tryptophan, tyrosine and lysine.[29] Only two alkaloid natural products, communesins G and H, are from low-temperature soil microorganisms. *P. rivulor*, a cold-active soil fungus, produces communesins G and H. These two compounds comprise polycyclic tryptamine molecules with an epoxide moiety and have alternating amide alkyl chains.[28] Unlike other communities isolated from marine habitats, these compounds did not exhibit any biological activity. However, these compounds might have some ecological role. Similarly, chanoclavine and elymoclavine ergoline alkaloids are produced by the psychrophilic soil fungus *Penicillium griseofulvin*, reviewed in a book chapter by Li et al.[30] However, no biological activity has been reported yet for these compounds.

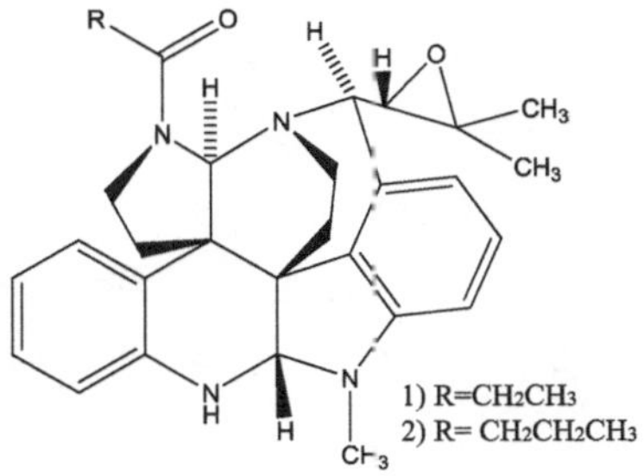

4. Communesins G (1) and H (2)

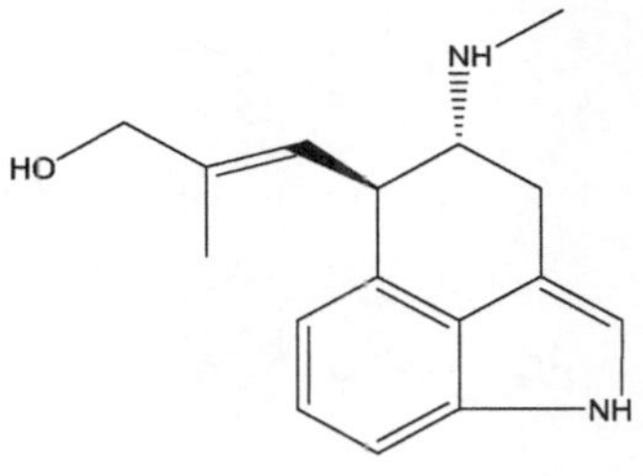

5. Chanoclavine I

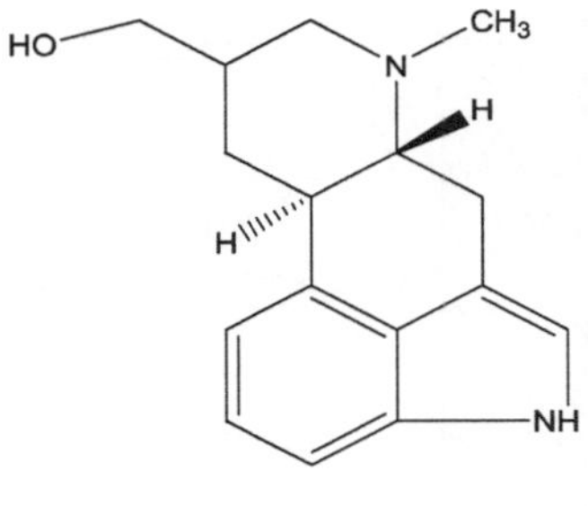

6. Elymoclavine

11.5.3 Asterric acid derivatives

Asterric acids are diphenyl ethers that play a significant role in treating early-stage idiopathic pulmonary fibrosis, act as endothelin receptor antagonists, and inhibit the vascular endothelial growth factor. However, several asterric acid derivatives have been declared very effective in treating renal failure and myocardial infarction. In 2008, five asterric acid derivatives comprising geomycins A–C, n-butyl asterrate and ethyl asterrate were isolated from a psychrotolerant *Geomyces* species from Fildes Peninsula and King George Island.[31] All three geomycins are different from each other. Geomycin A has a methyl ester moiety, while geomycins lack the methyl group, and geomycin C has a spirotricyclic moiety. Only geomycin B showed antifungal and antibacterial activity by efficiently inhibiting the growth of *Aspergillus fumigatus*, *Staphylococcus aureus* and *Escherichia coli*. This compound's carboxylic acid group was speculated to have antifungal activity, the 1st asterric acid bioactive derivative to have this activity.

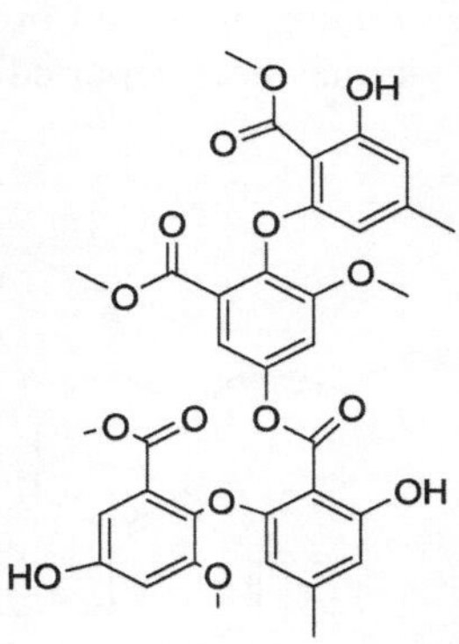

7. *Geomycins-A*

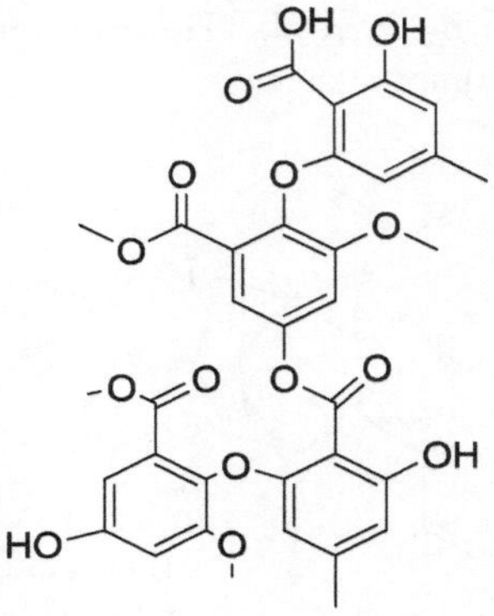

8. *Geomycins-B*

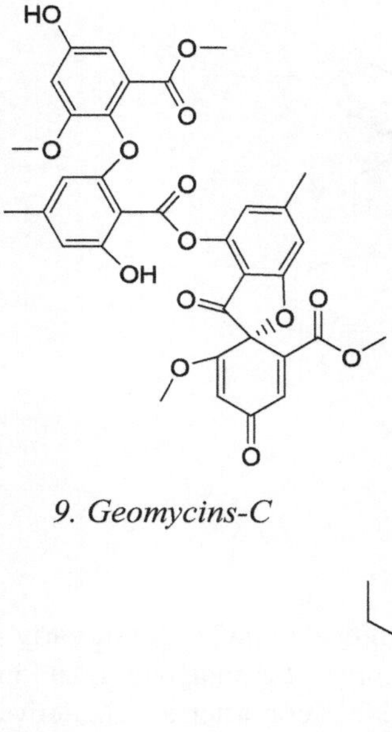

9. *Geomycins-C*

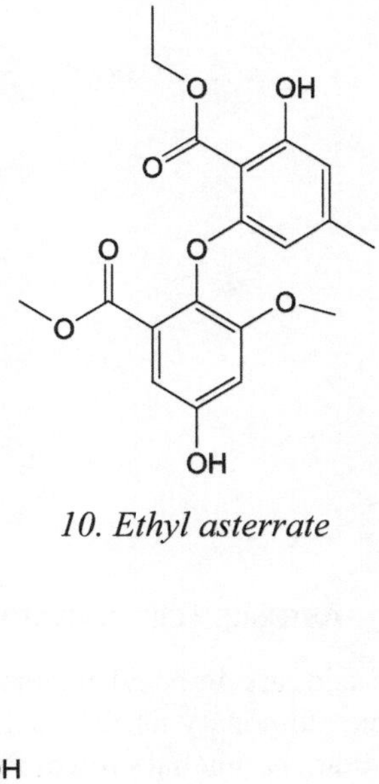

10. *Ethyl asterrate*

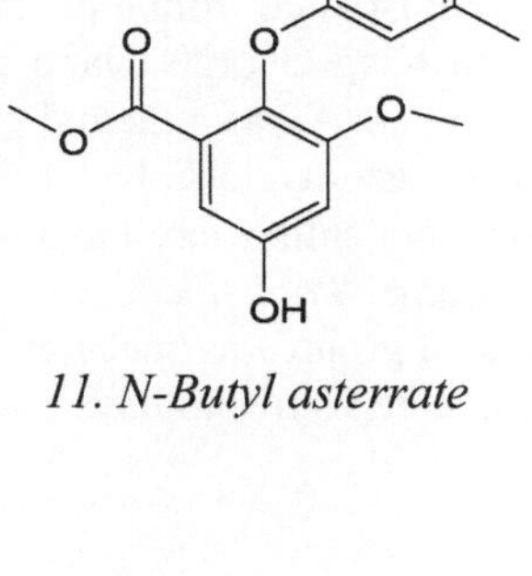

11. *N-Butyl asterrate*

11.5.4 Diketopiperazine and Epipolythiodioxopiperazines (ETPs)

Epipolythiodioxopiperazines (ETPs) are significant natural products having di- or polysulfide dioxopiperazine, a six-membered ring. These metabolites belong to fungal sources and have a broad range of bioactivities, like enzyme-inhibitory, immunosuppressive, antiviral, antitumour and antibacterial activities. To date, five new diketopiperazines, including oidioperazine A, oidioperazine B, oidioperazine C, oidioperazine D and chetracin D, and two new ETPs comprising chetracins B and C have been isolated from a psychrophilic soil fungus GW3-13 (*Oidiodendron truncatum*) isolated from Antarctic soil under lichen.[4] Among these, chetracin B shows anticancer activity against human hepatocellular adenocarcinoma Bel-7402, ovarian cancer cell lines (A2780), gastric cancer BGS-823, and colorectal cancer at very low concentration, with IC_{50} ranging from 3 to 28 nM. Similarly, chetracins C and D also displayed substantial toxicity in the cell, with IC_{50} in the range of 0.14–1.65 μM. Of these, chetracin C showed the more efficient cytotoxic activity. The broad bioactivities are due to the presence of the poly- or disulphide moiety. Removing these sulphide moieties or adding a reducing agent like dithiothreitol generally causes the complete loss of bioactivity.

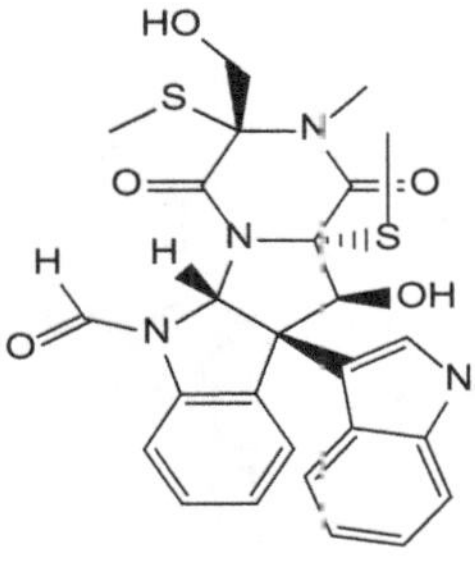

12. Oidioperazines A

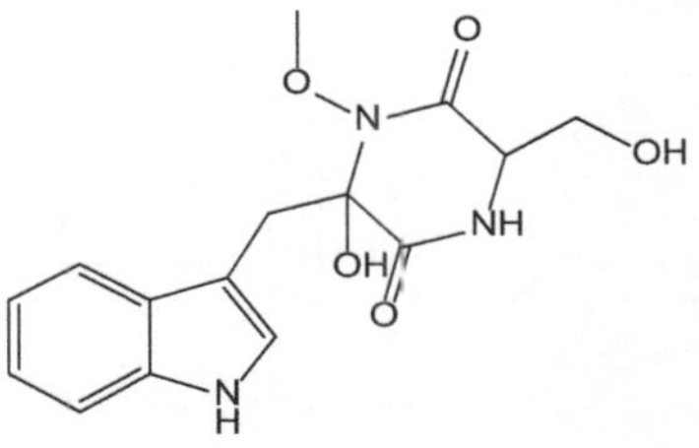

13. Oidioperazines B

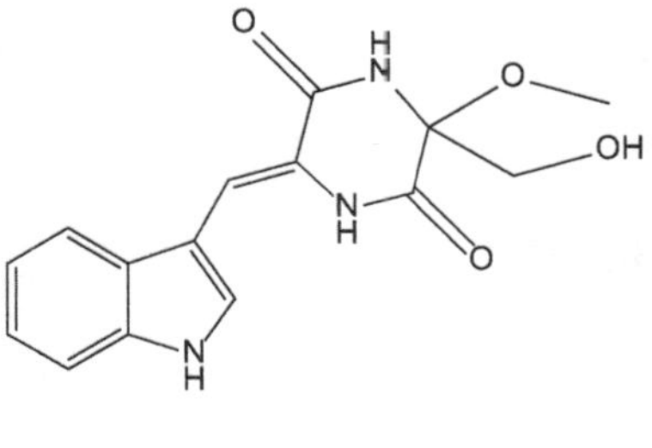

14. Oidioperazines C

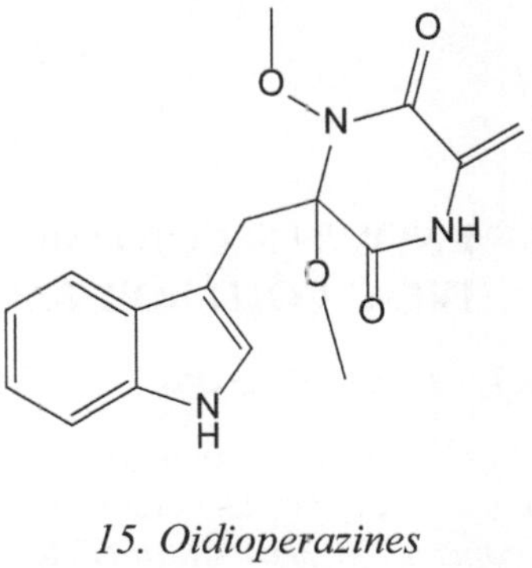

15. Oidioperazines

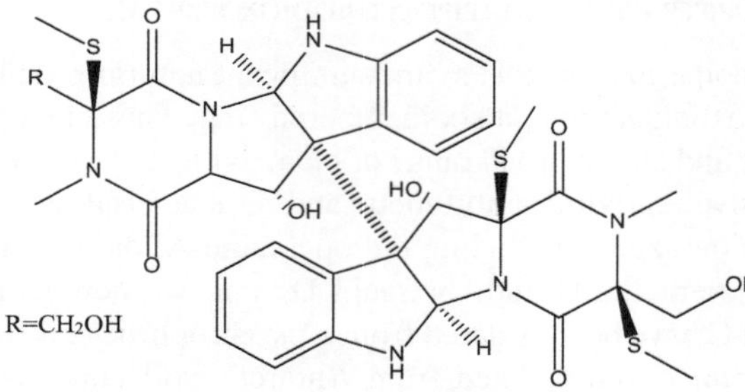

16. Oidioperazines

11.5.5 Amphotericin B

Amphotericin B is a Centers for Disease Control (CDC)-recognised natural product that exhibits antifungal and anti-leishmanial activity. Among cold soil microorganisms, *Penicillium nalgiovense* Laxa is a well-known efficient producer of amphotericin B.[32] This metabolite shows apparent activity against *Candida albicans, Escherichia coli* and *Streptococcus aureus.*

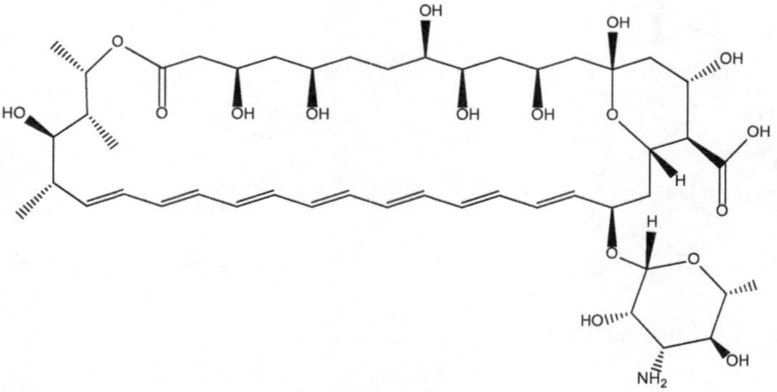

17. Amphotericin B

11.6 KNOWN SECONDARY METABOLITES FROM COLD SOIL BACTERIA

11.6.1 Angucyclinone

Angucyclinone is a polyketide (PKS) antibiotic encoded by PKS-II enzymes and consists of various natural products. Among cold soil microorganisms, only *Streptomyces griseus* strain NTK 97 produces a novel angucyclinone antibiotic known as frigocyclinone. This bacterium was isolated from Terra Nova Bay at Edmonson point, Antarctica and showed very efficient inhibition of Gram-positive bacteria.[19]

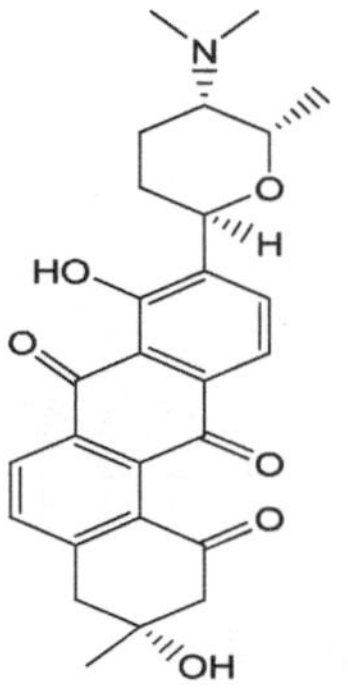

18. Frigocylclinone

11.6.2 Microcins

Microcins are extremely soluble, low-molecular-weight antibiotic substances (10,000 Da) made up of just a few amino acids formed ribosomally by a variety of Enterobacteriaceae strains with protease resistance and extreme pH and temperature tolerance. Serraticin A, a microcin developed by cold-active bacteria, has been identified in a soil bacterium closely similar to *Serratia proteamaculans* extracted from a sample of soil from Isla de Los Estados in Argentina. These were the first antimicrobial cold-active compounds showing activity against *Salmonella enterica* and *Escherichia coli.*[33] So, the possible mode of action is inhibition of the septation process or blocking the DNA.

11.6.3 Uncharacterised natural products from cold soil microorganisms

Soil is one of the most studied habitats for discovering bioactive secondary metabolites. Scientists have addressed the secondary metabolites from both fungal and bacterial species that are widespread and produce compounds with diverse biological activities. Fungi represent a significant source of novel secondary metabolites. The cold habitats, including the soil, are less explored. However, some studies reveal that the Ascomycota, particularly *Aspergillus* and *Penicillium* in fungi, abundantly prevail in cold soil with diverse biological activities. While among bacteria, members of the Proteobacteria and Actinobacteria are most prevalent, the Cyanobacteria and Firmicutes are also predominant but less frequent.[34, 35] The details of cold soil fungi and bacteria from cold soil with secondary metabolites are given in Table 11.1

TABLE 11.1
Bioactivity against selected microorganisms

Microbes	Genus	Species/Isolate	Sample site	Bioactivity	References
Fungi/Ascomycota	*Penicillium*	*rubens* 9496	Ellsworth Mountains, Antarctica	1	36
		allii-sativi 9451		2	
		allii-sativi 9458		1	
		allii-sativi 9508		2	
		brevicompactum 9446		2	
		brevicompactum 9448		1,2	
		chrysogenum 9534		2	
		chrysogenum 9466		1	
	Aspergillus	*sydowii* 9541	Ellsworth Mountains, Antarctica	1,3	
	Phoma	*herbarum* CCFEE 5015	Dry Valleys, Antarctica	3,4,5,6	37
		herbarum CFEE 5007		3,4,5,6	
		herbarum CCFEE 5020		3,4,5,6	
		herbarum CCFEE 459		3,4,5,6	
	Geomyces	sp. 2481	King George Island, Antarctica	2,5,7	

(*Continued*)

TABLE 11.1 (CONTINUED)
Bioactivity against selected microorganisms

Microbes	Genus	Species/Isolate	Sample site	Bioactivity	References
Bacteria/Actinomycetes	*Streptomyces*	SB47, SB33, SB81, SB72, SB9	Svalbard Islands	2,8,9,10,11,12,13,14,15	4
		PSY097	Signy Island	2,16	38
		NTK97	Terra Nova Bay	16,17	19
		5	East Antarctica	11,16,17,18,19	11, 39
		8		15,16,17,18,19,20,21,2 2,23,24,25	39
	Rhodococcus	2		26	
		RC56	Barrientos Island	2	4
	Arthrobacter	HPG8	East Antarctica	2,19,22,24,27	15
		HPH17		2,24,27	
		1		2,15,10,16,26	39
		9		2,15,26	
	Brevibacterium	BV34, BV35	Barrientos Island	15	4
		BV37		2,28	
	Nocardioides	*A-1*	East Antarctica	4,29,30,31	40
		ND52	Barrientos Island	15	4
	Micromonospora	*MM6*		28	
		MM32		32	
		18	East Antarctica	2,4,26	39
	Kocuria	*KC21*	Barrientos Island	2,28	4
	Janibacter	*JB26*		28	
	Lapillicoccus	*LC31*		15	
	Demetria	*DT40, DT41*		2	
	Gordonia	*G3, G4*		2,15	
Firmicutes	*Planococcus*	CHF8	East Antarctica	27,33,34	15
	Enterococcus	APR 210	Schirmacher Oasis	15,27,35,36,37	21
	Paenibacillus	PAMC26517	King George Island (Antarctica)	2,4,39	41

(*Continued*)

TABLE 11.1 (CONTINUED)
Bioactivity against selected microorganisms

Microbes	Genus	Species/Isolate	Sample site	Bioactivity	References
α-Proteobacteria	*Bradyrhizobium*	BR42	Barrientos Island	15	4
		BR45		15	
		BR62		15	
		BR65		15	
		BR82		15	
		BR88		15	
		BR96		15	
		BR105		15	
	Methylobacterium	MB20		15	
		MB104		15	
	Paracoccus	PC101		15	
	Sphingomonas	SM14		15	
β-Proteobacteria	*Janthinobacterium*	SMN33.6	Fildes Peninsula	5,32,36,40,41	42
γ-Proteobacteria	*Pseudomonas*	CG21	King George Island	5,38	43
		CrCD21		2, 33, 34	15
		WEK1		42	43
		MA2		42	
Bacteroidetes	*Pedobacter*	BG5		5,36,38,43,44	
Cyanobacteria	*Fischerella*		Haswell Island	4,45,46	40

1. *Cladosporium sphaerospermum*, 2. *Staphylococcus aureus*, 3. *Sarcina sp.* 4. *Bacillus subtilis* 5. *Escherichia coli*, 6. *Pseudomonas putida*, 7. *Aspergillus fumigatus*, 8. *Aeromonas salmonicida*, 9. *B. cereus*, 10. *Micrococcus luteus*, 11. *Ustilago nuda*, 12. *Vibrio vulnificus*, 13. *Yersinia enteroliti*, 14. *Trichosporon cutaneum*, 15. *Candida albicans*, 16. *Nematospora coryli*, 17. *Salmonella enterica*, 18. *Paecilomyces variotii*, 19. *P. vulgaris*, 20. *Vibrio anguillarum*, 21. *V. parahaemolyticus*, 22. *V. harveyi*, 23. *Penicillium notatum*, 24. *Proteus mirabilis*, 25. *Saccharomyces cerevisiae*, 26. *Sarcina lutea*, 27. *Listeria monocytogenes*, 28. *MRSA*, 29. *Micrococcus sp.*, 30. *Staphylococcus sp.*, 31. *Xanthomonas oryzae*, 32. *Pseudomonas aeruginosa* 33. *Brochothrix thermosphacta*, 34. *C. krusei*, 35. *Klebsiella pneumonia*, 36. *Listeria innocua*, 37. *Enterobacter cloacae*, 38. *Acinetobacter baumannii*, 39. *Serratia marcescens*, 40. *Vibrio parahaemolyticus*, 41. *Bacillus cereus*, 42. *Salmonella sp.*, 43. *Candida tropicalis*, 44. *Cladosporium cladosporioides*

11.7 CONCLUSION AND FUTURE PERSPECTIVE

Cold soil microorganisms, including fungi and bacteria, are incredible sources of secondary metabolites. These microorganisms face harsh environmental conditions, like extreme low temperatures and nutrient unavailability; however, these microorganisms adopt various mechanisms to cope with such conditions. Among these strategies is the production of antimicrobial compounds used as a weapon for interspecies competition, an essential survival strategy for inhabiting the cold environment. On the other hand, these natural products are active against various cell lines, bacteria and fungi. However, there is minimal data regarding natural products from low-temperature soil microorganisms – several compounds from these bacteria are found to be active against a wide variety of pathogenic microorganisms and cancer cell lines. However, a detailed investigation is required to explore the chemical diversity among these microorganisms. Another important aspect is that most microorganisms are not fully sequenced. To fully sequence them and identify the biosynthetic gene clusters and the attractive cryptic gene clusters, they can be cloned and expressed in the heterologous host either directly or via promoter engineering.

ACKNOWLEDGEMENT

We are thankful to Sohail Ahmad for proof reading and anonymous reviewers for reviewing to increase the quality of the manuscript.

REFERENCES

1. Margesin, Rosa, and Vanya Miteva. "Diversity and ecology of psychrophilic microorganisms." *Research in Microbiology* 162, no. 3 (2011): 346–361.
2. Piette, Florence, Salvino D'Amico, Gabriel Mazzucchelli, Antoine Danchin, Pierre Leprince, and Georges Feller. "Life in the cold: A proteomic study of cold-repressed proteins in the Antarctic bacterium *Pseudoalteromonas haloplanktis* TAC125." *Applied and Environmental Microbiology* 77, no. 11 (2011): 3881–3883.
3. Guglielmin, Mauro. "Advances in permafrost and periglacial research in Antarctica: A review." *Geomorphology* 155 (2012): 1–6.
4. Lee, Charles K., Béatrice A. Barbier, Eric M. Bottos, Ian R. McDonald, and Stephen Craig Cary. "The inter-valley soil comparative survey: The ecology of Dry Valley edaphic microbial communities." *The ISME Journal* 6, no. 5 (2012): 1046–1057.
5. Kääb, Andreas, Etienne Berthier, Christopher Nuth, Julie Gardelle, and Yves Arnaud. "Contrasting patterns of early twenty-first-century glacier mass change in the Himalayas." *Nature* 488, no. 7412 (2012): 495–498.
6. Inman, Mason. "Settling the Science on Himalayan Glaciers." *Nature Climate Change*, 1 no. 1003 (2010): 28–30.
7. Morita, Richard Y. "Psychrophilic bacteria." *Bacteriological Reviews* 39, no. 2 (1975): 144–167.
8. Helmke, Elisabeth, and H. Weyland. "Psychrophilic versus psychrotolerant bacteria-occurrence and significance in polar and temperate marine habitats." *Cellular and Molecular Biology* 50, no. 5 (2004): 553–561.

9. Rivkina, E. M., E. I. Friedmann, C. P. McKay, and D. A. Gilichinsky. "Metabolic activity of permafrost bacteria below the freezing point." *Applied and Environmental Microbiology* 66, no. 8 (2000): 3230–3233.
10. Bratchkova, Anna, and Veneta Ivanova. "Bioactive metabolites produced by microorganisms collected in Antarctica and the Arctic." *Biotechnology & Biotechnological Equipment* 25, no. supl (2011): 1–7.
11. Morgan-Kiss, Rachael M., John C. Priscu, Tessa Pocock, Loreta Gudynaite-Savitch, and Norman PA Huner. "Adaptation and acclimation of photosynthetic microorganisms to permanently cold environments." *Microbiology and Molecular Biology Reviews* 70, no. 1 (2006): 222–252.
12. Rodrigues, Debora F., Natalia Ivanova, Zhili He et al. "Architecture of thermal adaptation in an Exiguobacterium sibiricum strain isolated from 3 million year old permafrost: A genome and transcriptome approach." *BMC Genomics* 9, no. 1 (2008): 1–17.
13. Tehei, Moeava, and Giuseppe Zaccai. "Adaptation to extreme environments: Macromolecular dynamics in complex systems." *Biochimica et Biophysica Acta (BBA)-General Subjects* 1724, no. 3 (2005): 404–410.
14. O'Brien, Andrea, Richard Sharp, Nicholas J. Russell, and Sibel Roller. "Antarctic bacteria inhibit growth of food-borne microorganisms at low temperatures." *FEMS Microbiology Ecology* 48, no. 2 (2004): 157–167.
15. De Maayer, Pieter, Dominique Anderson, Craig Cary, and Don A. Cowan. "Some like it cold: Understanding the survival strategies of psychrophiles." *EMBO Reports* 15, no. 5 (2014): 508–517.
16. Rafiq, Muhammad, Noor Hassan, Maliha Rehman, and Fariha Hasan. "Adaptation mechanisms and applications of psychrophilic fungi." In *Fungi in Extreme Environments: Ecological Role And Biotechnological Significance*, pp. 157–174. Springer, Cham, 2019.
17. Giddings, Lesley-Ann, and David J. Newman. "Bioactive compounds from marine extremophiles." In *Bioactive Compounds From Marine Extremophiles*, pp. 1–124. Springer, Cham, 2015.
18. Bruntner, Christina, Tatiana Binder, Wasu Pathom-aree et al. "Frigocyclinone, a novel angucyclinone antibiotic produced by a Streptomyces griseus strain from Antarctica." *The Journal of Antibiotics* 58, no. 5 (2005): 346–349.
19. Sánchez, Leandro A., Fiorella F. Gómez, and Osvaldo D. Delgado. "Cold-adapted microorganisms as a source of new antimicrobials." *Extremophiles* 13, no. 1 (2009): 111–120.
20. Shekh, Raeesh M., P. Singh, S. M. Singh, and Utpal Roy. "Antifungal activity of Arctic and Antarctic bacteria isolates." *Polar Biology* 34, no. 1 (2011): 139–143.
21. Lerner, Phillip I. "Producing penicillin." *New England Journal of Medicine* 351, no. 6 (2004): 524–524.
22. Barriere, Steven L. "Clinical, economic and societal impact of antibiotic resistance." *Expert Opinion on Pharmacotherapy* 16, no. 2 (2015): 151–153.
23. Davies, Sally C., Tom Fowler, John Watson et al. "Annual Report of the Chief Medical Officer: Infection and the rise of antimicrobial resistance." *The Lancet* 381, no. 9878 (2013): 1606–1609.
24. Fair, Antibiotics. "Bacterial resistance in the 21st century, perspect." *Med. Chem* 6: 25.
25. Kanj, Souha S., and Zeina A. Kanafani. "Current concepts in antimicrobial therapy against resistant gram-negative organisms: Extended-spectrum β-lactamase–producing enterobacteriaceae, carbapenem-resistant enterobacteriaceae, and

multidrug-resistant Pseudomonas aeruginosa." In *Mayo Clinic Proceedings*, vol. 86, no. 3, pp. 250–259. Elsevier, 2011.
26. Dalsgaard, Petur W., John W. Blunt, Murray HG Munro, Thomas O. Larsen, and Carsten Christophersen. "Psychrophilin B and C: Cyclic Nitropeptides from the Psychrotolerant Fungus *Penicillium rivulum*." *Journal of Natural Products* 67, no. 11 (2004): 1950–1952.
27. Dalsgaard, Petur W., Thomas O. Larsen, and Carsten Christophersen. "Bioactive cyclic peptides from the psychrotolerant fungus *Penicillium algidum*." *The Journal of Antibiotics* 58, no. 2 (2005): 141–144.
28. Croteau, Rodney, Toni M. Kutchan, N. G. Lewis et al. "Biochemistry and molecular biology of plants." *American Society of Plant Physiologists*, Rockville, MD (2000): 1250–1318.
29. Lo Giudice, Angelina, and Renato Fani. "Antimicrobial potential of cold-adapted bacteria and fungi from polar regions." In *Biotechnology of Extremophiles*, pp. 83–115. Springer, Cham, 2016.
30. Li, Yan, Bingda Sun, Shuchun Liu et al. "Bioactive asterric acid derivatives from the Antarctic ascomycete fungus *Geomyces* sp." *Journal of Natural Products* 71, no. 9 (2008): 1643–1646.
31. Svahn, K. Stefan, Erja Chryssanthou, Björn Olsen et al. "*Penicillium nalgiovense* Laxa isolated from Antarctica is a new source of the antifungal metabolite amphotericin B." *Fungal Biology and Biotechnology* 2, no. 1 (2015): 1–8.
32. Svahn, K. Stefan, Erja Chryssanthou, Björn Olsen et al. "*Penicillium nalgiovense* Laxa isolated from Antarctica is a new source of the antifungal metabolite amphotericin B." *Fungal Biology and Biotechnology* 2, no. 1 (2015): 1–8.
33. da Silva, Thamar Holanda, Débora Amorim Saraiva Silva, Fábio Soares de Oliveira et al. "Diversity, distribution, and ecology of viable fungi in permafrost and active layer of Maritime Antarctica." *Extremophiles* 24, no. 4 (2020): 565–576.
34. Pearce, David A., Kevin K. Newsham, Michael AS Thorne et al. "Metagenomic analysis of a southern maritime Antarctic soil." *Frontiers in Microbiology* 3 (2012): 403.
35. Godinho, Valéria M., Vívian N. Gonçalves, Iara F. Santiago et al. "Diversity and bioprospection of fungal community present in oligotrophic soil of continental Antarctica." *Extremophiles* 19, no. 3 (2015): 585–596.
36. Onofri, Silvano, Massimiliano Fenice, Anna Rita Cicalini et al. "Ecology and biology of microfungi from Antarctic rocks and soils." *Italian Journal of Zoology* 67, no. S1 supl (2000): 163–167.
37. Yi Pan, Shing, G. Y. Tan, Peter Convey et al. "Diversity and bioactivity of actinomycetes from Signy Island terrestrial soils, maritime Antarctic." *Advances in Polar Science* 24, no. 4 (2013): 208–212.
38. Gesheva, Victoria. "Production of antibiotics and enzymes by soil microorganisms from the windmill islands region, Wilkes Land, East Antarctica." *Polar Biology* 33, no. 10 (2010): 1351–1357.
39. Gesheva, Victoria, and Evgenia Vasileva-Tonkova. "Production of enzymes and antimicrobial compounds by halophilic Antarctic Nocardioides sp. grown on different carbon sources." *World Journal of Microbiology and Biotechnology* 28, no. 5 (2012): 2069–2076.
40. Park, J-E., H-R. Kim, S-Y. Park, S-K. Choi, and S-H. Park. "Identification of the biosynthesis gene cluster for the novel lantibiotic paenilan from Paenibacillus polymyxa E681 and characterization of its product." *Journal of Applied Microbiology* 123, no. 5 (2017): 1133–1147.

41. Asencio, Geraldine, Paris Lavin, Karen Alegría et al. "Antibacterial activity of the Antarctic bacterium Janthinobacterium sp: SMN 33.6 against multi-resistant Gram-negative bacteria." *Electronic Journal of Biotechnology* 17, no. 1 (2014): 1–1.
42. Wong, Clemente Michael Vui Ling, Heng Keat Tam, Siti Aisyah Alias et al. "*Pseudomonas* and *Pedobacter* isolates from King George Island inhibited the growth of foodborne pathogens." *Polish Polar Research* 1 (2011).

12 Brass tacks of 'Omics' technologies for decoding soil microbiome

Richu Singla and Honey Goel*

CONTENTS

12.1 INTRODUCTION

Soil is a vital part of our bionetwork. If maintained naturally, it bears massive potential for encouraging the growth of microbial community, nutrient enrichment and contaminant deprivation. Soil microflora also plays a crucial role in plant–microbe communication through the decay of organic matter and maintaining the levels of the necessary nutrients through mineralisation. An apposite mechanism for the soil microbial systems needs to understand their types, their action and their interaction in the soil system. To make this soil microbial system more understandable, there are several contemporary genome-based

* Correspondence authors e-mail: richussingla@gmail.com

DOI: 10.1201/9781003354031-12

'Omics' techniques such as genomics, metagenomics, metaproteomics, metatranscriptomics, proteogenomics and metabolomics. Collectively, these techniques are able to explore the new vista to reveal these tiny creatures present in the soil ecosystem and its maintenance. Today, there is a sharp increase in the field of 'Omics' technology, which has improved the levels of understanding of soil microflora. 'Omics' refers to the quantitative characterisation of a set of molecules, especially biological in origin, for investigating the function, structure, physiology and molecular mechanisms of a set of organisms. As a result, recent 'Omics' technologies along with next-generation sequencing (NGS) may provide more consistent, efficient results to identify the soil microbiome and its interactions. Metagenomics has an extraordinary feature compared with conventional culture techniques: it may perceive a range of microorganisms even though they are not culturable. Metagenomics primarily aims at the study of organisms whose genetic materials can directly be assessed from the samples recovered from the environment, contrary to proteomics and transcriptomics, which are generally employed to explore the structure, function and components of different proteins and RNA molecules.

It is a matter of fact that these days, soils are predominantly contaminated due to the deposition of heavy metals in variable concentrations and indefinite compositions. The application of soil microflora and their mechanism of action to reinstate the contaminated sites has been the foremost aspect in the bioremediation of polluted soils.[1] Soil microbes serve as a vital source of naturally occurring enzymes and secondary metabolites such as antibiotics. Therefore, in order to surmount the obstacles in cultivation for the majority of the microbes, intensive efforts are being made by augmenting the time of incubation and properly simulating the naturally occurring environment along with the encapsulation of microbes, which allows them to form physically separate groups and continue their molecular exchanges with the environment.

12.2 DECIPHERING THE ENIGMA OF SOIL MICROBIAL COMMUNITY AND ITS NUTRIENT DYNAMICS VIA 'OMICS'

Soil has different characteristics and composite atmospheres, along with diverse microbiomes, which makes it very difficult to study. Many organisms are uncultured, which make it more difficult to study them at the molecular level. Therefore, metagenomics provides new insights to know the range of cultured as well as uncultured microorganisms along with their interaction and expression of genes. As a result, an association of genetic and functional similarity between the microbial communities can be established using 'Omics' technologies. A better assessment of the soil functionality along with its physiological characterisation can be executed using metagenomics, transcriptomics and proteomics in combination with diverse bioinformatics tools. With the advent of second-generation sequencing or NGS, the field of metagenomics is becoming more dependent on such technologies.

There, the role of 'Omics' technologies becomes indispensable to integrate the information along with metabolomics for a more accurate and precise assessment of the highly diversified heterogeneous microenvironment created by the soil microbiota. In a nutshell, although 'Omics' have revolutionised the detection and assessment of such microbial biota to a great extent, the current application of 'Omics' tools to the soil microbial community is in its infant phase, but with a proper bioinformatics approach and specific functional characterisation, it can mould the way to an exciting era of environmental microbiology.

12.3 APPLICATIONS OF 'OMICS' APPROACHES IN SOIL ECOSYSTEM

'Omics'-based information provides deep insight about the interaction of soil with microorganisms, which scrupulously conveys the details about the expression of a variety of genes, expressed proteins and metabolites. The end result of 'Omics' in the soil ecosystem is the appearance of novel data. 'Omics' is still in in its infant stage, and continuous improvement and research are progressing over time to explore the brass tacks of the soil microbial world along with its functions. In the last two decades, recent genomic/omic technologies have modernised the study of soil microbiology to a great extent and increased the scope of old conventional methods. Providentially, there are many complete genome sequencing projects being carried out globally, which are able to generate implausible data from sequencing for the study of soil microbiology, its activities and its function. Figure 12.1 depicts the different 'Omics' approaches for studying the soil microbiome.

12.3.1 THE RISE OF METAGENOMICS IN SOIL MICROBIOLOGY

Metagenomics does not need any prior knowledge of microorganisms; it gives the direct information about the genome analysis from the soil/environmental samples. Metagenomics discovers genes and expressed proteins as well as the complete genome sequence of uncultured bacterial entities, which is otherwise very difficult to understand. Metagenomics in soil gives the complete picture of the communication of the soil microflora with environmental factors and other microorganisms. It has been assumed that the species that are more dominating and expressed at a higher level than others have been enumerated from the soil primarily because they have the potential to grow on nutrient-rich media and in favourable temperature conditions. In earlier times, many uncultured organisms, such as eubacteria, have been distinguished and detected using phylogenetic analysis. Their DNA was extracted using two-dimensional agarose gel electrophoresis followed by cloning to a particular vector and identified through sequencing. But, the results were different from those previously reported; therefore, there was a need for more comprehensive and advanced genomic technique.[2] It has been noticed that more than 50% of the microorganisms in the environment cannot be cultured. In view of

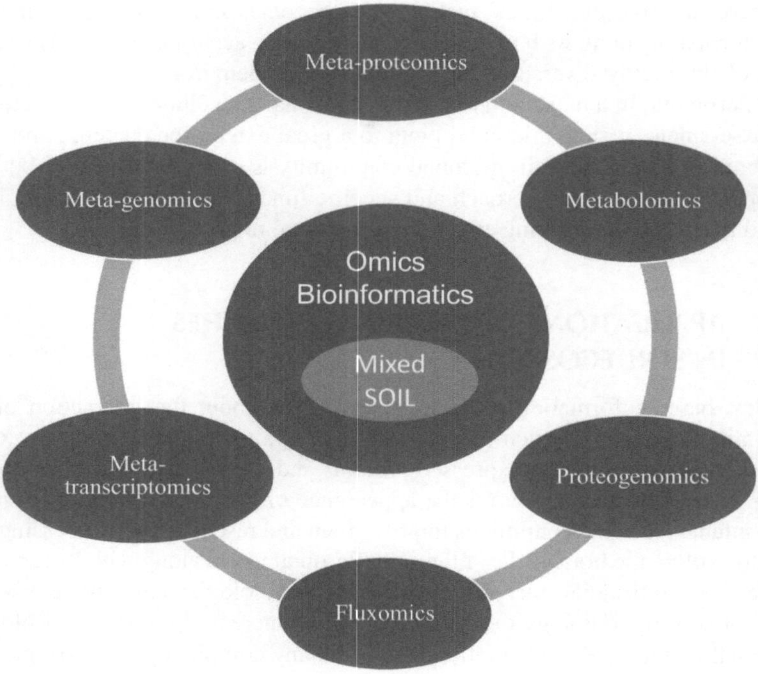

FIGURE 12.1 Multidimensional applications of a wide array of 'Omics' bioinformatics technologies in mixed soil.

this, the 16s rRNA technique modernised the whole concept to understand the whole ecosystem by unfolding the physiological functioning and biochemistry of the soil microorganisms.[3] Presently, shotgun sequencing techniques are used; these do not need to clone the extracted DNA to a vector but directly sequence the extracted DNA without forming any libraries. Shotgun sequencing in combination with NGS provides the elaborated data of the genome, which enables the metabolic/biochemical activities of environmental microorganisms to be checked. This combined technique also facilitates the identification of microorganisms in heterogeneous complex soil. Table 12.1 demonstrates the different soil microorganisms identified through the metagenomic approach. Table 12.2 overviews the comparative account of conventional approaches versus advanced technologies.

12.3.2 Emergence of Metatranscriptomics in Soil Microbiome

Metatranscriptomics studies target the gene expression of microorganisms, the response to a particular signal at a specific time and condition, and gene regulation in the natural environment by capturing the mRNA transcripts of microorganisms at a particular time and condition, which gives information about gene expression and its regulatory mechanisms. However, metatranscriptomics has the

TABLE 12.1
Study of diverse soil organisms/microbial community using Omics approaches

Advanced molecular technique	Target organism	Source	Conclusion of study	References
Genomics	*Pantoea agglomerans* strain	Moroccan vertisol	Colonisation of wheat roots	12
	Pseudaminobacter	French and Canadian agricultural soil	Characterisation of atrazine degradation genetically	13
	Soybean rhizobia	Alto Parana, Itapua, Paraguay	Characterisation of genetic material	14
	Archaea	Garden soil	Detection of uncultivated lineages	2
	Archaea	Sandy and forest soil	Genomic studies of archaea using large insert DNA libraries from soil	15
	Acidobacteria	Calcareous grassland soil	Evidence regarding environmental genomics	16
	Cyanidiales	Yellowstone National Park, Japan and New Zealand	Phylogenetic and biogeographic diversity	17
	Caulobacter	Oak Ridge, TN	Draft genome sequence of bacteria from less to heavy metals	18
	Populus deltoides	Fertile agricultural soil	Genomic analysis for root endophytic bacteria using single cell	19
Metagenomics	*Escherichia coli*	Jindong Valley, Korea	Production of indigo and indirubin in *E. coli* by description of a clone of forest soil metagenome	20
	Escherichia coli	Soil of forest	Expression of antifungal activity by using metagenome gene cluster	21
	Archaea, bacteria, viruses and fungi	Soil sample	Study of the genetic diversity in soil	22

(*Continued*)

TABLE 12.1 (CONTINUED)
Study of diverse soil organisms/microbial community using Omics approaches

Advanced molecular technique	Target organism	Source	Conclusion of study	References
Metatranscriptomics	*Geobacter* and *Methanothrix* species	Paddy soils	Interacts with methanogenic terrestrial environments by direct interspecies electron transfer	5
	Bacillus amyloliquefaciens	Axenic hydroponic cultures	Transcriptomic analysis of FZB42 in maize root exudates	23
	Soil microbial community	Tropical forest soil	Structural and functional analysis	24
	Prokaryotic bacteria	Waterlogged paddy soils	Identification of drivers of RNT in paddy soils	8
	Bacillus subtilis OKB105	Axenic hydroponic cultures	Transcriptome analysis of exposure in rice seedlings	25
	Pseudomonas aeruginosa	Manchester, UK	A transcriptional analysis of a phosphate-deficient *Lolium perenne* rhizosphere	26
Proteomics	*Rehmannia glutinosa*	Jiaozuo, Henan province, central China	Comparative study on monocultured rhizosphere soil	27
	Cupriavidus metallidurans	Quartz sand, kaolinite, montmorillonite and artificial soil	Effect of soil solid phases on the proteomic analysis	28
Metabolomics	Tobacco mosaic virus	Fertile soil	Investigation of infection in *Nicotiana tabacum* leaves	29
	Bacterial contamination	Garden soil	Analysis of the metabolomic response of *Brassica rapa* to preharvest contamination	30
	Different pathogens	Garden soil	Metabolic response of tomato leaves to various interactions	31
	Arabidopsis thaliana	Artificial medium with different Cd concentrations	Profiling of metabolites in response to cadmium exposure	32
	Sinorhizobium meliloti	Agricultural soil	Characterisation of acid tolerance by metabolomic approach	33

limitation that mRNA is unstable and has a short half-life, and it is challenging to recover the mRNA from microorganisms. Another factor is the lack of a poly-A tail in prokaryotic (bacterial and archaeal) mRNA transcripts, which are not able to synthesise cDNA. The latest advancement in metatranscriptomics has led to synthesising 'double RNA,' which is more stable, and both mRNA and rRNA are used to analyse the expression levels of microbial organisms. The diversity among microorganisms can be studied by directly sequencing the cDNA, which can compute the expressed gene profile.[4]

Consequently, recent 'Omics' approaches in molecular technologies have resulted in improved accuracy in characterising and classifying microorganisms. There is great multiplicity in the structure, metabolic and functional pathways of microorganisms. So, it becomes very difficult to evaluate the microbial communities quantitatively. Combined 'Omics' approaches should be pursued to interrogate the diversity between different microorganisms and the groups present in the soil microbiome.[5] Specific functions of soil eukaryotes can be studied by extracting the mRNA mixture having a poly-A tail directly from soil. Many extraction procedures are available for extracting the mRNA from soil samples, followed by converting the total RNA and mRNA to cDNA. Following this, cDNA libraries can be constructed by cloning into a suitable plasmid vector.[6] Then, finally, high-throughput sequencing can be performed to study the cDNA libraries, which provide information about gene expression at a particular time and condition. A combined technology of Illumina RNA sequencing of total RNA and pyrosequencing of rRNA genes was used for the identification of enzymes and their association with the lack of main components, like exoskeletons of arthropods, consisting of chitin, xylan, pectin and cellulose.[7] Reductive nitrogen transformation (RNT) gene transcript was identified in waterlogged paddy soil through the help of metatranscriptomics.[8]

Table 12.1 demonstrates the different soil microorganisms identified through a metatranscriptomics approach. However, metatranscriptomics has some limitations in view of the fact that microorganisms reveal a massive genetic range, the expression of proteins is high, and the microorganisms are present in mixtures. These limitations necessitate some advancement in these studies at protein level, i.e., metaproteomics. However, in a microbial mixture, identification and detection of all the proteins is highly problematic.[4]

12.3.3 Critical aspects of metaproteomics

Proteins mediate many of the significant actions of the soil microbiome. Therefore, metaproteomics has developed into the supremely important 'Omics' tool for determining the metabolic and physiological activities of soil microorganisms. Depending on the quality of the protein, its characterisation, identification and database design for analysis are a most important difficulty in the investigation of protein data. We can conquer this difficulty by using better techniques of protein extraction and creating a homogeneous database for easy protein characterisation. The combined approach of high-performance liquid chromatography (HPLC) and mass spectrometry (MS) has been used to recognise diverse metabolic functions of bacteria that are associated with plants and act as metal hyper-accumulators.[9] After the recognition of genera by 16S rRNA sequencing, an innate protein database was constructed for bacterial cells that can grow in a high-metallic environment. This proteomic database proved to be very significant for the identification of proteins in such environments. In comparison with metagenomics and metatranscriptomics, only a few proteins have been recognised to date.

The yield of proteins should be high for the suitable investigation and appropriate resolution of the soil for metaproteomics studies, and it becomes a really challenging task; extraction, isolation and purification, and finally, detection and identification of soil samples are critical steps. Metaproteomics in the soil microbiome has many challenges due to its complex nature, mixed protein content, the charismatic nature of microorganisms, and the presence of a variety of organic and inorganic compounds in the soil.[10] Many researchers have modified the extraction of proteins indirectly from the matrix of soil followed by cell concentration.[11] For that reason, more competent and consistent technologies are required for the isolation and identification of proteins and metaproteomics analysis that can conclusively define microorganisms' activity despite the heterogeneous nature of both the proteins and the soil matrix.

Therefore, a combination of metagenomics and metaproteomics tools is required to exploit various biocatalysts in soil samples collected from different environments. Functional metaproteomics serves as a proficient technique to quantify cell numbers or biomass of bacterial communities on an individual level. Therefore, the quantity of proteins becomes a measure of biomass quantification for assessing the composition of a microbial community. Hence, metaproteomics serves as a significant means of investigation for understanding the associations between microorganisms in an ecologically stable environment.

12.3.4 Applications of proteogenomics

Proteogenomics combines proteomics with metagenomics and transcriptomics for the better understanding of protein profiling. The fundamental standard of this technology is the processing of the same sample for the extraction of DNA and proteins. This combined technology is used for a much better understanding of the phylogenetic associations along with expression between diverse soil microorganisms.

Agarwal et al.[34] have used this powerful approach; proteogenomic (genomics and high-throughput proteomics) data of *Burkholderia* that were isolated from soils with a high uranium content revealed multiple mechanisms at the proteomic level. This technology provided breakthrough information on microbe interactions in a uranium environment, covering the physiology and metabolic activities of the microorganisms. Data was generated from the study of metal–microbe interactions using the proteogenomic (metagenomics and metaproteomics) data of *Microbacterium oleivorans* A9, an actinobacterium, with or without uranium under controlled conditions. The researchers uncovered 1532 proteins, and the presence or absence of uranium had a considerable impact on the expression levels of 591 of these.[35]

12.3.5 Applications of metabolomics

Metabolomics has various applications in ecosystem and molecular sciences, such as the reaction of microorganisms to the immediate environment, biotic factors such as temperature, infections and abiotic stress, and anthropogenic factors such as land development, overexploitation, species translocations and pollution control. Community-based or single-cell studies are carried out to establish the adaptability of a mixture of microorganisms related to diverse groups present in the soil using metabolites released by the bacteria. In the field of soil sciences, metabolomics plays an important role in monitoring disease, diagnosis, and risk measurement after exposure to a particular toxic substance. During the physiological, metabolic and biochemical processing of microorganisms, many metabolites are excreted, which may be the intermediate or end product and are naturally available in the cell. The characterisation, quantification and identification of metabolites are carried out by MS and nuclear magnetic resonance (NMR) spectroscopy, which produce characterised specific peaks. The pattern of each peak corresponds to a specific quantity and identity of the metabolite, followed by statistical analysis.[3] Other technologies, such as gas chromatography (GC) and HPLC, are also in use for the identification and quantification of metabolomes. Metabolomics may help in many aspects: a) it helps to understand the equilibrium of the soil or environment, as environment responds to any toxic environment; b) it helps to understand the regulatory mechanisms of microorganisms, as the expression of proteins changes with change in any culture activity; c) it is very significant in companion diagnostics, drug discovery and gene set analysis; d) this technology is very important to maintain a healthy environment.

Therefore, metabolomics plays an important role in the estimation of any ecological menace. Any unwanted materials, such as industrial waste products, pesticides or carcinogens, their mode of action and the possible risks related to these pollutants can be calculated in advance by metabolomics. The metabolic activities of microorganisms in soil after exposure to a variety of organic and inorganic pollutants/metals/toxic components can be studied, and some are depicted in Table 12.1.[37]

12.3.6 Application of fluxomics

Fluxomics is confined to the metabolic fluxes in a microorganism; at a particular time period, the concurrent flux of metabolites is measured through this 'Omics' technique.[36] This flux of metabolites, also termed the *fluxome*, provides an incorporated functional output of the cellular technology and an improved way to link a microorganism's response to any changes in its genome and adjacent environment. Metabolic fluxes are excreted in very low amounts in comparison

TABLE 12.2
Comparison of conventional approaches versus advanced technologies

Sr. No.	Conventional technology	Advanced technology
1	Less sensitive	Highly sensitive
2	Less efficient	More efficient
3	More labour required	Less labour required
4	Low throughput	High throughput
5	One sample in one go	More samples in one go
6	Simple procedure	Complex procedure but for knowledge
7	Does not assess risk to human health	Assesses risk to human health
8	Use of very easy and simple software for the processing of generated data results in less informative records	Highly sophisticated software for the processing of generated data results in informative records

to the genes and proteins normally present in the cell. Fluxomics is very close to metabolomics. It consists of subcellular trafficking (transporters, vesicles) linked with a metabolic network of enzymatic reactions connecting metabolites, which aims to determine rates of reactions and design a flux map that translates enzymatic activities and regulation. Therefore, fluxomics is considered advantageous in comparison to genomics or proteomics.

12.4 COMPLEXITIES AND CHALLENGES IN 'OMICS'

The 'Omics' approach proved to be an advanced approach in soil microbiology, but still, 'Omics' in the soil microbiome exhibits myriad limitations. Any transient variations further augment the difficulty of 'Omics' study. Another chief concern in the isolation of nucleic acids is the partial cell lysis, which ultimately affects the nucleic acid yield of the soil microbiome.[38] Many organic compounds, such as humic acid, present in soil also hinder polymerase chain reaction (PCR) amplification. During PCR amplification, short PCR products are amplified and sequenced, which may produce small contigs during assembly and give hardly any significant information. A large amount of data is sometimes generated, which requires expertise in bioinformatics, and any wrong information generated may create problems in analysing future data, too.[39]

12.5 CONCLUSION

With the expansion and relevance of modern 'Omics' tools, the field of soil ecology is undergoing unusual alterations. Post-genomic 'Omics' technologies allow us to cross-examine the structural and functional variety of the soil microbiome, and researchers have smashed the genetic, biochemical and metabolic variation of microflora present in this environment. There are many questions yet to be answered, such as: a) can we count the microbial species present on the Earth? b) can we judge the metabolic activity of the microorganisms? c) is it possible to see the microbial interactions/associations in the soil ecosystem? d) how are microorganisms regulated by biological, chemical and physical factors? e) is it possible to understand the microbial response to any stress factor? Many of these questions remain to be answered. Understanding uncultured microorganisms is still a discouraging assignment, as genes characterised through metagenomics have no representatives in earlier databases and are yet to be identified as new species/genes. Even though substantial improvement in the identification and characterisation of the soil microbiome has been achieved by recent 'Omics' technologies, many chapters still remain to open. Inter-disciplinary coordination should be implemented for 'Omics' technologies to divulge the linkage between gene profiling, protein profiling, and response to physiological changes and environmental conditions in order to provide new insight into exploiting the soil microbiome. The advancement of multiple 'Omics' approaches should be the main concern in the coming years, involving researchers such as microbiologists, molecular biologists and biostatisticians with appropriate expertise and training in their respective fields.

In future, hard work will be needed to link the association of environmental activity with predator–prey interaction and to understand the magnitude of not only bacterial populations but also viral contributions, which are available in the soil, but their study is still in its infancy.[40]

ACKNOWLEDGEMENTS

There are no acknowledgements.

REFERENCES

1. Sharma, Pooja, Surendra Pratap Singh, Hafiz MN Iqbal, and Yen Wah Tong. "Omics approaches in bioremediation of environmental contaminants: An integrated approach for environmental safety and sustainability." *Environmental Research* 211 (2022): 113102.
2. Quaiser, Achim, Torsten Ochsenreiter, Hans-Peter Klenk et al. "First insight into the genome of an uncultivated crenarchaeote from soil." *Environmental Microbiology* 4, no. 10 (2002): 603–611.
3. Biswas, Rimi, and Angana Sarkar. "'Omics' tools in soil microbiology: The state of the art." *Advances in Soil Microbiology: Recent Trends and Future Prospects* (2018): 35–64.
4. Simon, Carola, and Rolf Daniel. "Metagenomic analyses: past and future trends." *Applied and Environmental Microbiology* 77, no. 4 (2011): 1153–1161.
5. Rotaru, Amelia-Elena, Pravin Malla Shrestha, Fanghua Liu et al. "Direct interspecies electron transfer between Geobacter metallireducens and Methanosarcina barkeri." *Applied and Environmental Microbiology* 80, no. 15 (2014): 4599–4605.

6. Yadav, Rajiv K., Claudia Bragalini, Laurence Fraissinet-Tachet et al. "Metatranscriptomics of soil eukaryotic communities." In *Microbial Environmental Genomics (MEG)*, pp. 273–287. Humana Press, New York, 2016.
7. Ivanova, Anastasia A., Carl-Eric Wegner, Yongkyu Kim et al. "Identification of microbial populations driving biopolymer degradation in acidic peatlands by metatranscriptomic analysis." *Molecular Ecology* 25, no. 19 (2016): 4818–4835.
8. Masuda, Yoko, Hideomi Itoh, Yutaka Shiratori et al. "Predominant but previously-overlooked prokaryotic drivers of reductive nitrogen transformation in paddy soils, revealed by metatranscriptomics." *Microbes and Environments* 32, no. 2 (2017): 180–183.
9. Mattarozzi, Monica, Marcello Manfredi, Barbara Montanini et al. "A metaproteomic approach dissecting major bacterial functions in the rhizosphere of plants living in serpentine soil." *Analytical and Bioanalytical Chemistry* 409, no. 9 (2017): 2327–2339.
10. Nannipieri, Paolo. "Role of stabilised enzymes in microbial ecology and enzyme extraction from soil with potential applications in soil proteomics." In *Nucleic Acids and Proteins in Soil*, pp. 75–94. Springer, Berlin, Heidelberg, 2006.
11. Shrestha, Him K., Manasa R. Appidi, Manuel I. Villalobos Solis et al. "Metaproteomics reveals insights into microbial structure, interactions, and dynamic regulation in defined communities as they respond to environmental disturbance." *BMC Microbiology* 21, no. 1 (2021): 1–17.
12. Amellal, N., G. Burtin, F. Bartoli, and T. Heulin. "Colonization of wheat roots by an exopolysaccharide-producing Pantoea agglomerans strain and its effect on rhizosphere soil aggregation." *Applied and Environmental Microbiology* 64, no. 10 (1998): 3740–3747.
13. Amellal, N., G. Burtin, F. Bartoli, and T. Heulin. "Colonization of wheat roots by an exopolysaccharide-producing Pantoea agglomerans strain and its effect on rhizosphere soil aggregation." *Applied and Environmental Microbiology* 64, no. 10 (1998): 3740–3747.
14. Chen, Lu Shi, Antonio Figueredo, Fábio O. Pedrosa, and Mariangela Hungria. "Genetic characterization of soybean rhizobia in Paraguay." *Applied and Environmental Microbiology* 66, no. 11 (2000): 5099–5103.
15. Treusch, Alexander H., Arnulf Kletzin, Guenter Raddatz et al. "Characterization of large-insert DNA libraries from soil for environmental genomic studies of Archaea." *Environmental Microbiology* 6, no. 9 (2004): 970–980.
16. Quaiser, Achim, Torsten Ochsenreiter, Christa Lanz et al. "Acidobacteria form a coherent but highly diverse group within the bacterial domain: Evidence from environmental genomics." *Molecular Microbiology* 50, no. 2 (2003): 563–575.
17. Toplin, J. A., T. B. Norris, C. R. Lehr et al. "Biogeographic and phylogenetic diversity of thermoacidophilic cyanidiales in Yellowstone National Park, Japan, and New Zealand." *Applied and Environmental Microbiology* 74, no. 9 (2008): 2822–2833.
18. Utturkar, Sagar M., Annette Bollmann, Ryann M. Brzoska et al. "Draft genome sequence for Caulobacter sp. strain OR37, a bacterium tolerant to heavy metals." *Genome Announcements* 1, no. 3 (2013): e00322-13.
19. Utturkar, Sagar M., W. Nathan Cude, Michael S. Robeson et al. "Enrichment of root endophytic bacteria from Populus deltoides and single-cell-genomics analysis." *Applied and Environmental Microbiology* 82, no. 18 (2016): 5698–5708.
20. Lim, He Kyoung, Eu Jin Chung, Jin-Cheol Kim et al. "Characterization of a forest soil metagenome clone that confers indirubin and indigo production on *Escherichia coli*." *Applied and Environmental Microbiology* 71, no. 12 (2005): 7768–7777.

21. Chung, Eu Jin, He Kyoung Lim, Jin-Cheol Kim et al. "Forest soil metagenome gene cluster involved in antifungal activity expression in Escherichia coli." *Applied and Environmental Microbiology* 74, no. 3 (2008): 723–730.
22. Fierer, Noah, Mya Breitbart, James Nulton et al. "Metagenomic and small-subunit rRNA analyses reveal the genetic diversity of bacteria, archaea, fungi, and viruses in soil." *Applied and Environmental Microbiology* 73, no. 21 (2007): 7059–7066.
23. Fan, Ben, Lilia C. Carvalhais, Anke Becker et al. "Transcriptomic profiling of *Bacillus amyloliquefaciens* FZB42 in response to maize root exudates." *BMC Microbiology* 12, no. 1 (2012): 1–13.
24. Urich, Tim, and Christa Schleper. "The 'double-RNA' approach to simultaneously assess the structure and function of a soil microbial community." *Handbook of Molecular Microbial Ecology I: Metagenomics and Complementary Approaches* (2011): 587–596.
25. Xie, Shanshan, Huijun Wu, Lina Chen et al. "Transcriptome profiling of *Bacillus subtilis* OKB105 in response to rice seedlings." *BMC Microbiology* 15, no. 1 (2015): 1–14.
26. Zyśko, Agnieszka, Hervé Sanguin, Andrew Hayes et al. "Transcriptional response of Pseudomonas aeruginosa to a phosphate-deficient *Lolium perenne* rhizosphere." *Plant and Soil* 359, no. 1 (2012): 25–44.
27. Wu, Linkun, Haibin Wang, Zhixing Zhang et al. "Comparative metaproteomic analysis on consecutively Rehmannia glutinosa-monocultured rhizosphere soil." *Plos One* 6, no. 5 (2011): e20611.
28. Giagnoni, Laura, Francesca Magherini, Loretta Landi et al. "Soil solid phases effects on the proteomic analysis of Cupriavidus metallidurans CH34." *Biology and Fertility of Soils* 48, no. 4 (2012): 425–433.
29. Choi, Young Hae, Hye Kyong Kim, Huub JM Linthorst et al. "NMR metabolomics to revisit the tobacco mosaic virus infection in Nicotiana t abacum leaves." *Journal of Natural Products* 69, no. 5 (2006): 742–748.
30. Jahangir, Muhammad, Ibrahim Bayoumi Abdel-Farid, Young Hae Choi, and Robert Verpoorte. "Metal ion-inducing metabolite accumulation in *Brassica rapa*." *Journal of Plant Physiology* 165, no. 14 (2008): 1429–1437.
31. López-Gresa, M. Pilar, Federica Maltese et al. "Metabolic response of tomato leaves upon different plant–pathogen interactions." *Phytochemical Analysis: An International Journal of Plant Chemical and Biochemical Techniques* 21, no. 1 (2010): 89–94.
32. Sun, Xiumei, Jixiu Zhang, Haijun Zhang et al. "The responses of *Arabidopsis thaliana* to cadmium exposure explored via metabolite profiling." *Chemosphere* 78, no. 7 (2010): 840–845.
33. Draghi, Walter Omar, María Florencia Del Papa, Aiko Barsch et al. "A metabolomic approach to characterize the acid-tolerance response in *Sinorhizobium meliloti*." *Metabolomics* 13, no. 6 (2017): 1–12.
34. Agarwal, Meenakshi, Ashish Pathak, Rajesh Singh Rathore et al. "Proteogenomic analysis of Burkholderia species strains 25 and 46 isolated from uraniferous soils reveals multiple mechanisms to cope with uranium stress." *Cells* 7, no. 12 (2018): 269.
35. Gallois, Nicolas, Béatrice Alpha-Bazin, Philippe Ortet et al. "Proteogenomic insights into uranium tolerance of a Chernobyl's Microbacterium bacterial isolate." *Journal of Proteomics* 177 (2018): 148–157.
36. Wiechert, Wolfgang, Oliver Schweissgut, Hitomi Takanaga, and Wolf B. Frommer. "Fluxomics: Mass spectrometry versus quantitative imaging." *Current Opinion in Plant Biology* 10, no. 3 (2007): 323–330.

37. Myrold, David O, and Paolo Nannipieri. "Classical Techniques Versus Omics Approaches." (2013).
38. Delmont, Tom O., Patrick Robe, Sébastien Cecillon et al. "Accessing the soil metagenome for studies of microbial diversity." *Applied and Environmental Microbiology* 77, no. 4 (2011): 1315–1324.
39. Vinaixa, Maria, Emma L. Schymanski, Steffen Neumann et al. "Mass spectral databases for LC/MS-and GC/MS-based metabolomics: State of the field and future prospects." *TrAC Trends in Analytical Chemistry* 78 (2016): 23–35.
40. Merges, Dominik, Alexandra Schmidt, Imke Schmitt et al. "Metatranscriptomics reveals contrasting effects of elevation on the activity of bacteria and bacterial viruses in soil." *Molecular Ecology* (2022).

13 Ecological role of psychrophiles and mechanism of adaptation in various cold ecozones

Osikemekha Anthony Anani, Mohd. Shahnawaz, Puja Gupta, Osayomwanbo Osarenotor and Frances Ngozi Olisaka*

CONTENTS

* Corresponding author e-mail: osikemekha.anani@edouniversity.edu.ng; cybert2004@gmail.com

DOI: 10.1201/9781003354031-13

13.1 INTRODUCTION

Microorganisms (viruses, Eukarya, Archaea and Bacteria) play a significant role in the recycling of nutrients and are considered as important indicators of a well-balanced and unperturbed ecosystem.[1] Microorganisms having the natural ability to conquer extreme environmental conditions like cold are termed *psychrophiles*, from the Greek: *psukhros* (cold) and *philein* (love).[1,2] The ability to overcome bio-geographic limitations, responses to different abiotic and biotic parameters, and tropism (lifestyle) are other features psychrophiles also possess to conquer their immediate environment.[1]

Psychrophiles can tolerate a wide range of environmental temperatures below 30 °C because as temperature rises, the rate of enzymatic activities increases, which leads to kinetic heat effects, cellular activities and the swift growth of the microorganisms.[1,2] However, several studies have reported tolerances of microbes in varied temperatures, e.g., *Methanococcoides burtonii* and *Methanogenium frigidum* (15 °C, 18 °C, 23 °C, 28 °C),[1] *Chusquea subtessellata* (4–25 °C),[3] *Sphingopyxis alaskensis* and haloarchaea (4–10 °C–20 °C).[4,5]

As per studies, climate change poses a serious biological threat to animal diversity, distribution, forced migration, loss and biological change as well as ecological threats: cold, heat, nutrient deficiencies, drought and salinity.[6,7] Microbiomes and microbial symbionts impact negatively or positively on their host(s), which in turn has been found to modify the ecosystem severely. This can be seen in the relationship between the severity and incidence of diseases caused by microbes in plants.[8,9] Some studies[10–12] have recounted the impacts of global warming occasioned by climate change on the rise of sea temperatures, which has impacted *Symbiodinium*, the algal coral endosymbiont, although not the bacterial form. In addition, plants with this form of association have been shown to possess a high level of resistance and adaptation toward drought-abiotic conditions.

Global warming impacts on biodiversity have recently revealed how microbes can either acclimatise or adapt to severe conditions or migrate from them. Some studies have reported[13–15] how symbionts of microbes and mycorrhiza conquer their immediate environment (high temperature) and suggested that psychrotrophic fungi can thrive better at 10 °C and 20 °C.

In the present chapter, an attempt will be made to discuss the ecological role and adaptive mechanism of psychrophiles in various cold ecozones in the following aspects: (1) to highlight the ecological role and mode of action of psychrophiles, (2) to list the genes used by psychrophiles to adapt in the cold environment, (3) to discuss the biotechnological potential of psychrophiles and (4) to overview the metagenomic properties of psychrophiles in soil.

13.2 PSYCHROPHILES: THEIR ECOLOGICAL ROLE AND MODE OF ACTION

D'Amico et al.[2] evaluated the challenges faced by the psychrophiles to proliferate and survive at very low environmental temperatures. Psychrophiles' adaptability in various cold ecozones is portrayed in Figure 13.1.

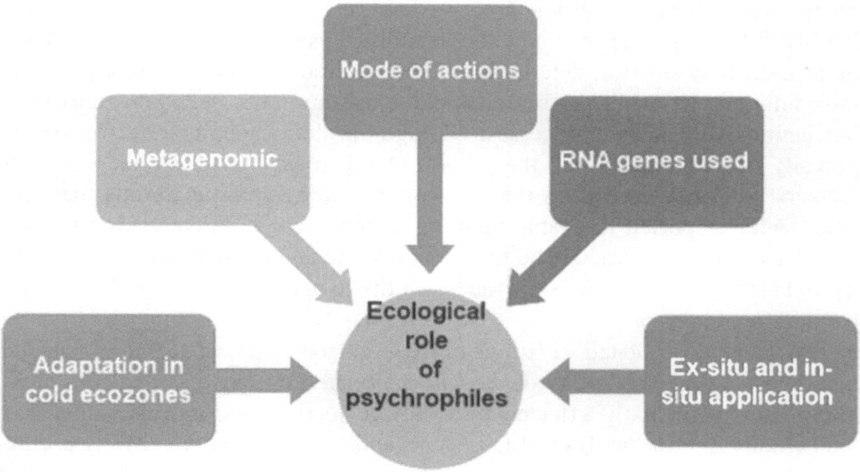

FIGURE 13.1 Ecological role of psychrophiles and mechanism of adaptation in various cold ecozones.

Psychrophiles thrive in extreme abiotic conditions – cold, even at 0 °C – in the presence of active enzymes, which aid in the sustenance of their biological cell cycle.[16] Feller[16] showed that temperature dependence is usually at a low state during the cold period, and the catalytic action is reduced moderately.

Cavicchioli et al.[17] stated that about 90% of the biosphere is covered with water (temperature ≤5 °C), which sustains several microbial diversities. These water bodies span the polar, alpine and deep-sea regions, ecosystems that are always cold and harbour psychrophiles. A study[18] evaluated the survival tactics used by psychrophiles to cope with the severe cold environment. Ecological factors like soil-microbial physiological actions affect the soil composition and structure, and a decrease in environmental temperature is one of the leading consequences of commercial plant losses globally.[19] However, severe cold ecosystems have been reported to be a good habitat for psychrotolerant and psychrophilic PSB (phosphate solubilising bacteria) that can tolerate several ranges of low-temperature settings while still keeping their metabolic activities intact. The production and expression of proteins that have antifreeze properties and stress-induced DNA segments at reduced temperatures encourage their survival instinct during severe cold stress. Psychrophiles can elicit plant development and growth by providing phosphorus and essential nutrients to soil that is highly deficient, which is a new functional feature of psychrophilic PSB. Plants require rhizosphere bacteria while growing under intense or severe environmental conditions. By contrast, it aids to improve their biological performance. In addition, the utilisation of psychrophilic PSB mixtures in the form of formulations has been reported to be effective in the optimisation of plants in low-temperature settings.[19] Psychrophiles are also adapted to the cold environment by using biomolecules like enzymes and proteins from cell factories characterised by heat lability.[20] These properties make it essential for psychrophiles' biomolecules to be utilised in cosmetics, detergents, industrial feed and food technologies, and molecular and biomedical research. Cold temperature has been known to affect biochemical and nucleic processes, folding of proteins, the rate of diffusion and porosity of membranes, and the growth of organisms. Psychrophiles are conformers with many enzymes that possess crystallographic materials that have been widely reported to enable them to adapt to the cold. These structures or materials are also characterised to have moderate and low-temperature specificity and stability, which improves the flexibility to prevent freezing effects during temperature changes.[21]

Hassan et al.[22] isolated 42 strains of bacteria from a glacier region in Pakistan and reported that the fluidity of the *Chryseobacterium frigidisoli* PB4T's membrane was significantly affected by exposure to the environment with varied temperature. For the analysis of fatty acids and methyl esters, GC-MS (gas chromatography/mass spectrometry) was used to characterise them by the liberation of the acid-catalysed methanolysis. Gram –ve and Gram +ve groups were cultivated at temperatures 5, 15, 25 and 35 °C to ascertain the effect of temperature on the fatty acid and cell membrane distribution, like the br-FAs and n-MUFAs (branched fatty acids and straight-chain monounsaturated fatty acids), which accounted for over 70% of the total analysed fatty acid components.

13.3 HIGHLIGHTS OF THE VARIOUS RNA GENES USED BY THE PSYCHROPHILES TO ADAPT TO THEIR ENVIRONMENT

Psychrophiles acclimatise to cold environments through certain adaptations, e.g., production of cold shock proteins, RNA chaperones, RNA helicases, oxidative and osmotic stress, synthesis of carotenoids and transformation factors, and peptidoglycan and membrane modifications in compatible solutes.[23] The presence of these genes is not unique to psychrophiles, but an abundance of such genes has been reported in microbes confronting low temperatures. The above-mentioned adaptations are discussed briefly in the following.

13.3.1 Cold shock proteins

Bacteria contain numerous cold shock protein (CSP) gene homologs, e.g., ribonuclease, RNA helicase (RNase, DEAD-box) and CSP.[8] It is worth stating that not all CSP genes are induced by cold.[23] CspI, G, CspE, B and CspA of *Escherichia coli* are among the few examples that are stimulated by cold.[24] An extremely preserved biological component – the nucleic acid acting as a bond region known as the cold shock domain (CSD) – is found in CSP.[25] CSD has two ribo-nucleoprotein binding subjects that aid its binding to bull's eye DNA and RNA.[26] CSP avoids the establishment of hairpin structures, thereby weakening two-degree structures in bull's eye RNA at a reduced temperature, and enables transcription and translation at low temperatures.[27] CSPs were initially identified as vital cold-induced proteins involved in tolerance to cold stress and playing an important role in controlling a wide variety of gene expressions and physiological responses in bacteria. However, it has been hypothesised that a few of them participate in several cellular functions that support healthy growth and stress adaptation. It has been demonstrated that a few CSPs also promote host cell invasion and ethanol, osmosis, oxidation, pH, starvation and stress tolerance.

The effects of heat and cold shock on psychrophilic lactic acid microbes were studied by Duru.[28] The cspA (cold shock protein A) gene was noted as the major protein in the genes of the microbes. Besides, cshB and cshBA, the DEAD-box ribonucleic acid helicase genes, were also reported to play an important part in the response of the psychrotrophic lactic acid bacteria, while the DEAD-box ribonucleic acid helicase contributes to the unwinding and degradation of the ribosome and RNA biosynthesis all through the cold shock.[28] Some of the CSPs detected in mesophiles are expressed constitutively in psychrophiles as cold-acclimation proteins (CAPs) to adapt to extremely low temperatures.[2]

13.3.2 RNA chaperones

Chaperones aid in the efficient folding of proteins. Low temperature appears to elevate the synthesis of RNA chaperones.[4] Cold-adapted chaperones such as DnaK,[29] GroEL[30] and the molecular chaperone IbpA[31] when expressed in *E. coli* increase the mesophilic bacterium's ability to grow at low temperatures. Co-chaperonin GroES, chaperonin GroEL, molecular chaperone DnaK, molecular chaperone IbpA, molecular chaperone HSP90 and molecular chaperone GrpE were noted among the chaperones elicited at reduced temperature.[32]

The TF (ribosome-bound-trigger factor) assists in protein folding.[33] The vast majority of nascent polypeptides generated by the ribosome interact with TF during co-translational processes. TF assists in the folding of the majority of proteins (about 70% of all proteins) quickly after synthesis. Interestingly, the remaining chaperones were noted to be heat shock proteins (HSPs), but TF is a cold shock protein in *E. coli*.[34] Cold-adapted bacteria have been found to overexpress TF at low temperatures, whereas HSP chaperones are downregulated.[35] The downregulation of chaperones' HSP in the cold implies that the majority of these are produced at transitorily higher ambient temperatures.[35] Certain bacterial species produce the chaperonins Cpn10 60, which aid in their adaptation to the evironment.[36] The chaperone proteins Hsc66 and TF played an important part in the cold shock response,[37] with the presence of other proteins important for trehalose, cellular metabolism and the biological manufacturing of lipids.

A study on the TRiC chaperonin in the psychrophilic yeast *Glaciozyma antarctica* revealed that the genes are consistently produced irrespective of cold or heat shock. TRiC chaperonin is known to transport denatured luciferase back to its active state at low temperatures.[38] Reduced ion pair electrostatic interactions are required for efficient protein folding by such chaperones. Besides, non-polar amino acids like isoleucine, phenylalanine, leucine, tryptophan and cysteine are involved in hydrophobic residues and are modified in the TRiC and TCP1-1 ring chaperonin and complex subunits, respectively, as an adaptation to low temperatures. The hydrophobic residues are substituted with hydrophilic residues like glutamine, histidine, lysine, aspartic acid, arginine and serine, increasing the residue distance, which eventually assists in cold adaptation; e.g., isoleucine is substituted for valine at TRiCa at location 372 in *G. antarctica*. The isoleucine is put into hydrophobic compartments formed by the distanced residues. Interactions with aromatic amino acids are also involved that aid in cold adaptation.[38] A few psychrophile genomes are reported to have the gene for proline, an osmolyte that is known to support the cell's viscosity and maintain structural integrity. A higher concentration of lysine is observed in the genomes of psychrophiles, which retains protein flexibility at reduced temperatures.[39]

13.3.3 RNA helicases

RNA helicases are mostly associated with abiotic stress. Low temperature is among the most studied biotic stresses. RNA helicases are engaged in the establishment of cold-adapted ribosomes and RNA degradosomes.[40] Such helicases are reported to be important for the weakening of RNA and DNA secondary assemblies.[2] They might help to slow down the RNA secondary assemblies for effective transformation in the cold.[41] SrmB and CsdA are among the DEAD-box RNA helicase family members that are encoded by *E. coli*. These appear to be involved in the adaptation of cells to the cold.[42] SrmB contributes to the biosynthesis of ribosomes. CsdA is more strongly required at low temperatures than SrmB. The deletion of the csdA gene drastically affects growth at low temperatures.[43] Numerous biological processes, including mRNA degradation, translation initiation and ribosome synthesis, have been attributed to CsdA. It is involved in mRNA degradation, which is the primary mechanism through which it contributes to cold acclimatisation. CsdA's helicase activity is essential at low temperatures. It was demonstrated that helicase-deficient CsdA mutants do not complement cold sensitivity.[42] The RNA helicases have also been seen to be overexpressed in many psychrophiles, like *Exiguobacterium sibiricum*,[44] *Psychrobacter arcticus arcticus*,[45] *Pseudoalteromonas haloplanktis*, *Sphingopyxis alaskensis*[4] and *Shewanella denitrificans*[46], at low temperature.[47] These underpin that the microbes have evolved adaptive strategies at reduced temperatures by the production of proteins. Therefore, RNA helicases increase the interactive crosstalk between helicase and protein cofactors, modifying the arrangement of functional protein complexes under various environmental conditions.[40]

13.3.4 Oxidative stress and osmotic stress

Increased heat and respiration in cold conditions give rise to oxidative stress. This results in the creation of the catalase-peroxidase (KatG) and the biogenesis of the proline use protein (PutA). As a result of these molecules, H_2O_2 synthesis gradually increases, activating proteins that inhibit reactive oxygen species (ROS).[36] Some other examples of molecules that are produced in response to OS (oxidative stress) are β-lactamase superfam II, catalase peroxiredoxin, glyoxylase or related hydrolase, glutathione peroxidase, spermidine synthase, thioredoxin reductase, ABC glycine/proline-betaine messenger, ATPase constituent, choline glycine-betaine transporter, choline-trehalose-6-phosphate synthase, osmoprotectant binding protein, trehalose-6-phosphatase and maltooligosyl trehalose synthase. Na^+/proline-symporter, Na^{+}/H^{+}-antiporters dehydrogenase or related flavoprotein are some of the osmotic regulators that are also produced in response to oxidative stress. Genes related to osmotic stress like vicR and vicK are thought to be in control for coping with ecological pressure; resD and resE for oxygen stress; narGHIJ and nreABC operon genes for respiration; and ompR and envZ for osmotic operon and upshift genes osmVW and osmXY.[48]

Gas solubility and harmful ROS generation both greatly increase at low temperatures. The presence of many genes encoding dismutases, superoxidase and catalases in *Colwellia psychrerythraea* and *Pseudoalteromonas haloplanktis* has improved the antioxidant capability to combat harsh environmental conditions.[2] LCPFAs (long-chain polyunsaturated fatty acids) like eicosapentaenoic, docosahexaenoic and arachidonic acid are formed at enhanced rates at lower temperatures.[49] Long-chain polyunsaturated fatty acids defend the cell membrane against ROS, thereby functioning as a shield from OS at a reduced temperature.[50]

13.3.5 Carotenoid synthesis enzymes

Psychrophiles synthesise carotenoid pigments in response to fluctuating temperatures to preserve the fluidity and viscosity of the cell-protective structure/membrane.[18] At very low temperatures, the cell membrane is stabilised by the carotenoids. The carotenoid biosynthetic pathways in psychrophiles contain genes such as crtB, idi, crtZ, crtY and crtI (phytoene synthase, isopentenyl-diphosphate delta isomerase, beta-carotene hydroxylase, lycopene beta cyclase and phytoene dehydrogenase).[39] Non-polar and polar carotenoid pigments are formed by a variety of Antarctic microbes to control membrane fluidity and help maintain homeoviscosity during temperature variations.[44] Many research groups have co-related the presence of carotenoids with the cold adaptation of microbes.[51] These pigments are known to serve a variety of important functions: light harvesting (photosynthesis by microbes), photoprotection (against UV rays) and antimicrobial.

13.3.6 Transcription and translation

Psychrophilic microorganisms adapt to low temperatures by controlling the activity of certain enzymes, e.g., peptidyl-prolyl *cis-trans* isomerase, extension factor and RNA polymerase. These enzymes are shown to retain action at extremely low temperatures in several psychrophilic microorganisms. Protein-folding rates may need to be maintained at low temperatures through overexpression at these temperatures.

Previous investigations have demonstrated that microbes control translation at low temperatures through the translation initiation factors (IF-3, IF-2 and IF-1).[26] Initiation factors IF-3 greatly favour CSP translation in *E. coli*.[52] Other examples of translation factors that are regulated are tRNA-dihydrouridine synthase, threonyl carbamoyl adenosine dehydratase (tRNA A37), GTPase translation initiation factors IF-2 and IF-3, GTPase translation elongation factor EF-Tu and EF-G, transcription-blocking protein NusA, transcription-related element helicase of the NusB superfamily II RNA and DNA, SNF2 superfamily II RNA helicase, and RNA or DNA helicase.[32]

13.3.7 Membrane and peptidoglycan modifications

Certain microbes enhance their peptidoglycan production, which eventually thickens the peptidoglycan layer, allowing it to withstand low temperatures. For instance, *Planococcus halocryophilus* has an unusual process that results in a thickened outer cell wall made of peptidoglycan, choline and calcium carbonate.[53] Some microbes additionally have LPS (lipopolysaccharide) structures in the outside layer of the cell to adapt to the cold habitat. Short-chain unsaturated fatty acids included in the LPS component further increase the fluidity in cold settings. Recent transcriptome research revealed that some organisms had larger quantities of outer-layer proteins and glycosyltransferases.[54] Benforte et al.[55] demonstrated impaired growth at reduced temperatures in an Antarctic microbe due to a transmutation in the glycosyltransferase wapH gene of LPS. It resulted in a decreased ability of Antarctic bacteria to thrive at cold temperatures. Other examples of proteins engaged in membrane modification are glycosyltransferases, reductase, 3-oxoacyl-[acyl-carrier-protein]-synthase III (KASIII) and 3-oxoacyl-[acyl-carrier-protein]. Lower growth temperatures also result in a complex content of polyunsaturated and unsaturated anteiso-branched fatty acids, *cis*-unsaturated double-bonds, acyl-chain and methyl-branched fatty acids.[2] By adding steric limitations that alter the packing order or decrease the number of contacts in the membrane, this changed composition is expected to play a crucial part in improving membrane fluidity. In addition, increasing levels of the big pigments of non-polar carotenoids, proteins and lipid groups contribute to increased fluidity of membrane structures.[56]

13.3.8 Compatible solutes and related compounds

Osmoprotection and cryoprotection are significantly aided by the presence of certain solutes, e.g., betaine, glycine, glycerol, mannitol, sucrose, sorbitol and trehalose. Additionally, these solutes act as suppliers of carbon, nitrogen and energy.[57] Compatible solutes stabilise cellular membranes at cold temperatures, lower the cytoplasm's freezing point, and stop and scavenge free radicals and macromolecule accumulation.[58] Nunn et al.[59] studied the genome of *Colwellia psychrerythraea* and reported the genes that code for proteins involved in the degradation and synthesis of polymers, nitrogen and polyamides. Such genes can be viewed as a possible special adaptation to cold conditions.[59] Ghobakhlou et al.[60] evaluated the metabolomic scrutiny of the Arctic strain N33 (*Mesorhizobium* sp.) and reported elevated accumulation of valine, threonine and sarcosine when cultivated at 4 °C. The microbe was found to behave like a cryoprotectant.

Many bacteria that are native to the cold ecosystem are found to manufacture PHAs (polyhydroxyalkanoates). These are stand-in polymers with significant physiological functions. They operate as an active sink of carbon, thus accumulating and reducing imbalanced growth conditions, nitrogen, and other macro- or micronutrients.[61] PHAs have ecological significance and give bacteria increased survival and resilience to a range of environmental stressors.[62] The DNA analysis of *C. psychrerythraea* showed the capacity to produce PHA, which is connected to synthesising and breaking down fatty acids.[59] Several gene duplications in the families of enoyl-CoA hydratase and acyl-CoA dehydrogenase were found, likely suggesting the diversity of the PHAs that can be produced.[57] The capacity of freshwater alpha-, beta- and gammaproteobacterial isolates from Antarctica to synthesise PHA was thought to be a common trait at early sites.[62] PHB synthesis was discovered to be essential for freezing and cold survival and growth, respectively.[62] PHB accretion boosted movement and prolonged planktonic cells' ability to survive in the bacteria's cold-adapted biofilms, indicating that having the ability to accumulate PHB may provide an adaptive benefit for attachment in those environments.[62]

13.3.9 Antifreeze proteins (AFPs)

The AFPs are a structurally varied set of proteins that are crucial in preventing the production of intracellular and extracellular ice crystals, which would otherwise cause cell death.[63] These proteins prefer ice and can lower a solution's freezing point without changing the temperature of the solution's melting point.[63] The process, termed TH (thermal hysteresis), also changes the ice shape foundation, reliant on the protein binding site and the level of ice formation.[64] The antifreeze and ice-binding proteins allow the cells of microbes to thrive in severe cold settings.[48] AFPs play distinct functions in various organisms. These function by either preventing freezing or increasing tolerance to freezing. Mostly, AFPs attach to frozen crystals, impede their growth, and help to prevent ice formation and re-crystallisation, which keeps the organism from freezing. According to Davies et al.,[65] the AFPs isolated from microorganisms display a typical triangular helical fold paired with a stretch of °-helical assembly. The previously indicated pattern differs noticeably from AFPs produced by other organisms.[65] Gilbert et al.[66] showed that the Antarctic lake bacterium *Marinomonas primoryensis* contains AFPs.[66]

Muryoi et al.[67] reported antifreeze proteins from rhizobacterium strain GR12-2 (*Pseudomonas putida*) in Arctic plants.[67] Nine AFP-encoding genes were reported by several research teams working on *G. antarctica*. The gene expression analysis showed that each AFP has a range of expression levels contingent on the environment, temperature and organism.[63] It's interesting to note that just two AFP-encoding genes were found in the expressed sequence tags (EST) studies, one transcript for AFP7 (GaAFP7) being found in the −12 °C library and the other for AFP9 (GaAFP9) being found in the 0 °C library. Additionally, quantitative PCR (qPCR) analysis was used to support the data and revealed similar expression levels.[67]

13.4 BIOTECHNOLOGICAL APPLICATION OF PSYCHROPHILES

Margesin and Feller[20] examined the use of psychrophiles for environmental purposes. Low environmental temperature is common in temperate parts of the world. Organisms like microbes that live at this temperature can tolerate and conquer their immediate habitat. Such organisms can be employed for the manufacture of biological cell factories or biological molecules for the remediation of wastewater and pollutants *ex situ*.[20] These biological molecules include mostly enzymes and proteins with good catalytic actions and definite heat lability, which have been found important and useful in industrial applications in different areas like cosmetics, detergents, feed and food technologies, and medical and biological research. Hamid et al.[68] reviewed the potential use of psychrophilic yeast for biotechnology. About 7% of the global area is covered with cold ecological systems, which are colonised by microbes that are cold tolerant. Psychrophilic microorganisms like yeast are important assets in such an ecosystem because they have the potential to adapt to a very low temperature due to their varied physiological activities that could generate cold shock proteins and enzymes. So, psychrophilic yeast could be a potential source for biotechnological use for the production of industrial food, chemicals, pharmaceutical and medical products, and textiles. They can also be used for ecological applications to manage environmental stress in plants. Hamid et al.[68] recommend further studies on enzymes and cold proteins for better applications in other fields like bioengineering.

The microbes that inhabit the frozen or cold environment possess exopolysaccharides, polyunsaturated lipids, and proteins in their membranes that enable them to conquer the environment successfully.[69] Psychrophiles have a great pool of cold enzymes that enable them to tolerate a wide range of environmental temperatures. Due to their biotechnological potential, cold enzymes could be employed in a wide range of environmental and industrial uses, like the textile, domestic, agricultural, cosmetic, biosensor and bioremediation industries, and in the manufacture of enzymes and pharmaceutical products.[69] In addition, the components of their enzymes could also be employed in distant satellite planets monitoring to explore the possible existence of life therein through the evidence of the presence of gases. The biopotential of psychrophiles in the food and manufacturing industry using cold enzymes like β-mannanases, lipases, pullulanases, xylanases, amylases, proteases, pectinase and β-galactosidase is highly recommended.[70] These enzymes are used to improve the quality of certain catalytic effects in food production processes. Biocatalysts or enzymes sourced from microbes have been stated to possess potential use for industrial applications in the waste management, pharmaceutical, chemical, agriculture and food sectors.[71] The enzymes from psychrophiles possess strong biochemical efficiency that will help to lower the time of reaction of the substrate and the inputs of energy in the industrial process. The process is cheap, non-toxic and eco-friendly, and the use of such biological enzymes from microbes in the industrial production of food has also highlighted the need for collaboration between industry and academia for the production of an engineered biocatalyst that has the potential to take part in a reaction without altering its initial process.[71]

13.5 SOIL PSYCHROPHILES' METAGENOMIC PROPERTIES

Low available nutrient contents, low annual precipitation, extreme temperatures, low soil moisture and freeze-thaw cycles are all factors that limit microbial activity in Arctic soils.[72] The Bacteroidetes and Proteobacteria are the main taxa in all metagenomes involving psychrophiles in the soil of Antarctic lakes.[73] In studying the consortia of microbes, various CASPs, CIPs and C-RSGs (cold-associated general stress-responsive proteins, cold-induced proteins and cold-responsive stress genes) in soil, sediment and Antarctic lake metagenomes were also studied. It was observed that in all samples investigated, *Gloeobacter*, *Haliangium*, *Anaeromyxobacter* and *Myxococcus* were dominant in the soil and lake sediment metagenomes. However, genes for exopolysaccharide biosynthesis were dominant in Lake Untersee soil metagenomes. In conclusion, it was found that although different consortia of microbes are found in many metagenomes, they share similar C-RSGs needed for their sustenance and survival in severe Antarctic settings. In Finnish Lapland, a study[74] reported that Gram-negative bacteria (Alpha-, Beta- and Gammaproteobacteria phyla) are dominant, and 60% of all isolates from Arctic tundra soils are Pseudomonads, while the phylum Acidobacteria was rich in soils having low pH [75]

The Actinobacteria, Bacteroidetes and Proteobacteria were also reported as the predominant phylogenetic groups of the glacier ice on topsoil in Northern Schneeferner, Germany,[76] with some important genera like *Stenotrophomonas*, *Cryobacterium*, *Sphingomonas* and *Polaromonas*. After comparison of all pyrosequencing-derived sequences of efficient diversity of the Northern Schneeferner glacier, it was observed that some microbial species were reported to have similar homologies, such as dicarboxylate metabolism, glyoxylate, sucrose, starch, butanoate propanoate and pyruvate. However, some of the homologous sequence genes take part in the phosphate-pentose cycle, citrate cycle and gluconeogenesis/glycolysis pathway. It was observed that the genes responsible for growth specifically under anaerobic settings were detected when the microbe utilised nitrite and/or nitrate as an electron acceptor. In addition, several enzymes and compounds essential for a psychrophilic lifestyle and adaptation to living in a low-nutrient environment were also detected. It has also been recorded that in cold soil environments, ammonia oxidisers are known to be present.[77] Shen et al.,[78] in a metagenomic data comparison, investigated the linkage between physiological and genomic characteristics of the psychrophilic bacteria Arthrobacter. The analyses showed that the Arthrobacter family comprises a clade with associates having genomic characteristics with a strong capacity to reproduce faster due to the possession of an amino acid that is projected to reduce the rigidity of a great number of proteins by improving enzyme action at reduced temperatures. Together, the genomic and physiological traits compel adaptation for a psychrophilic clade of Arthrobacter. The metagenomics of bacterial communities in Antarctica, using the CRISPR spacer content, revealed that the genus *Flavobacterium* was a major bacterium in all sampling sites,[79] whereas another study[80] on the cold desert soils reported *Acinetobacter* and *Streptomycetes* as major phylogenetic groups. A group of researchers suggested that soil nutrients, global warming/climate change and reduced atmospheric pressure have resulted in a significant decline in biodiversity.[81] As a consequence of this, the bacterial and physio-chemical properties of soil within cold regions of the world, like the cold desert of the Himalaya, face severe ecological stress that prompts an environmental and biological strategy to survive. In soil at low altitudes and extremely cold conditions, *Firmicutes* and *Bacteroidetes* were dominant, while *Bacteroides*, *Flavobacterium* and *Cytophaga* were found in large quantities at the altitude. A similar study by Joshi et al.[82] revealed the presence of the species *Pseudomonas helmanticensis*, *Pseudomonas mandelii*, *Brevibacillus invocatus* and *Arthrobacter humicola* in the high altitude of the western Indian Himalaya. Actinobacteria were reported to be the major phylum in the Antarctic tundra soil, and their synergistic relationship with the genera *Streptomyces*, *Streptosporangium* and *Amycolatopsis* was also documented.[83] These microbes also possess cellular and cold-active catabolic enzymes with potential for catabolic breakdown of lignocellulose.[83]

13.6 PSYCHROPHILES' ADAPTABILITY IN VARIOUS COLD ECOZONES

The mechanisms used by psychrophilic PSB in cold ecozones to conquer their environment include enzymes, nucleic acids, phospholipids and phytin.[19] For example, organisms like *Pseudomonas azotoformans*, *Pseudomonas proteolytica*, *Pseudomonas palleroniana* and *Pseudomonas aeruginosa.* can grow

TABLE 13.1
Psychrophiles' adaptability in various cold ecozones

Sr. No.	Species	Type of acids used for adaptability	Ecozone(s)
1	Fluorescent *Pseudomonas* sp.	Malic, citric, formic, succinic, lactic, 2-ketogluconic, oxalic and gluconic acids	The Himalayan cold desert[86]
2	Strain BIHB 723 (*Acinetobacteria rhizosphaerae*)	Formic, malic, lactic, 2-ketogluconic, oxalic and gluconic acids	The trans-Himalayan cold desert[87]
3	Strain BIHB 783 (*Ralmella* sp.)	Iso-citric, citric and gluconic acids	*Hippophae rhamnoides* rhizosphere[88]
4	*Pseudomonas* sp.	Gluconic acid	In samples of glacial ice[89]
5	Strain AZ17 (*Bacillus* sp.) and strain AZ5 (*Pseudomonas* sp)	Lactic, citric, acetic, gluconic and oxalic acids	The rhizosphere of chickpea[90]
6	*Bacilli* sp.	Propionic, succinic, malic, citric, lactic and gluconic acids	Phosphate rocks in mine region and rhizospheres of wheat plants[91]
7	*Enterobacter*, *Serratia*, *Pseudomonas* and *Pantoea*	Fumaric, succinic, gluconic, citric and oxalic acids	Rhizospheres of wheat plants[92]
8	*Paenibacillus* sp., *Burkholderia* and *Bacillus*	Acetic, formic, succinic, tartaric, citric, oxalic and gluconic acids	Paddy rice field[93]
9	*Serratia plymuthica*	Acetic acid	Soils[94]
10	*Pseudomonas azotoformans*, *Pseudomonas proteolytica*, *Pseudomonas paleroniana* and *Pseudomonas* sp.	Succinic, citric, malic and lactic acids	High-altitude region of Himalayan soil – Indian[84]

and survive at the temperature range between 15 and 25 ° by the production of malic and oxalic acids in the wild, while laboratory-cultured organisms released succinic, citric and lactic acids (Table 13.1).[19,53] In addition, enzymes released by their membranes, such as GED (glucose dehydrogenase), can mediate the released acids, such as gluconic acid, which is an important solubilisation effect in soil microbiomes.[85] The adaptive features are possibly linked to the study of the metagenomics, ecozones, climate warming surrounding the microbes, and the ecological factors as well as the genes responsible for their tolerance limits towards self-preservation and conservation of their innate ability (Figure 13. 2).

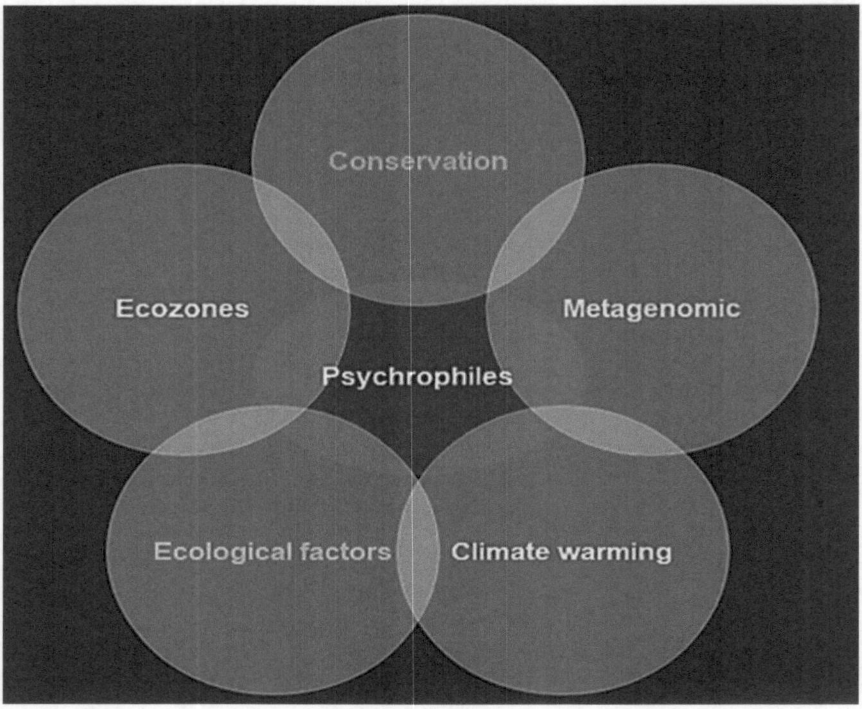

FIGURE 13.2 Strategies for psychrophiles' survival in cold ecozones.

13.7 CONCLUSIONS

Based on the preceding discussion, the following conclusions can be drawn:

- Psychrophiles have a natural ability to conquer extreme environmental conditions like cold. As the temperature rises, the rate of enzymic activities increases, which leads to kinetic heat effects, cellular activities and the swift growth of the microorganisms.
- The mechanism of action and the ecological role of psychrophiles show that they can proliferate and survive at very low environmental temperatures.
- Psychrophiles acclimatise to cold environments through certain adaptations, e.g., production of cold shock proteins, RNA chaperones, RNA helicases, oxidative and osmotic stress, synthesis of carotenoids, transformation factors, peptidoglycan and membrane modifications in compatible solutes.
- The cold enzymes could be employed in a wide range of environmental and industrial uses, such as the textile, domestic, agricultural, cosmetic,

biosensor and bioremediation industries and in the manufacture of enzymes and pharmaceutical products.

- Low available nutrient contents, low annual precipitation, extreme temperatures, low soil moisture and freeze-thaw cycles are all factors that limit microbial activity in Arctic soils.

ACKNOWLEDGEMENT

The authors thank their various institutions for allowing them to carry out this review work.

REFERENCES

1. Cavicchioli, R. "Commentary on the concept of a psychrophile". *The ISME Journal* 10, (2016): 793–795.
2. D'Amico, Salvino, Tony Collins, Jean-Claude Marx et al. "Psychrophilic microorganisms: Challenges for life." *EMBO Reports* 7, no. 4 (2006): 385–389.
3. Seas, Carolina, and Priscila Chaverri. "Response of psychrophilic plant endosymbionts to experimental temperature increase." *Royal Society Open Science* 7, no. 12 (2020): 201405.
4. Ting, Lily, Timothy J. Williams, Mark J. Cowley et al. "Cold adaptation in the marine bacterium, Sphingopyxis *alaskensis*, assessed using quantitative proteomics." *Environmental Microbiology* 12, no. 10 (2010): 2658–2676.
5. DeMaere, Matthew Z., Timothy J. Williams, Michelle A. Allen et al. "High level of intergenera gene exchange shapes the evolution of haloarchaea in an isolated Antarctic lake." *Proceedings of the National Academy of Sciences* 110, no. 42 (2013): 16939–16944.
6. Steidinger, Brian S., Thomas W. Crowther, Jingjing Liang et al. "Climatic controls of decomposition drive the global biogeography of forest-tree symbioses." *Nature* 569, no. 7756 (2019): 404–408.
7. Wieczynski, Daniel J., Brad Boyle, Vanessa Buzzard et al. "Climate shapes and shifts functional biodiversity in forests worldwide." *Proceedings of the National Academy of Sciences* 116, no. 2 (2019): 587–592.
8. Zhang, Yan, David H. Burkhardt, Silvi Rouskin et al. "A stress response that monitors and regulates mRNA structure is central to cold shock adaptation." *Molecular Cell* 70, no. 2 (2018): 274–286.
9. Hofstetter, Richard W., and Fernando E. Vega, eds. *Bark Beetles: Biology and Ecology of Native and Invasive Species*. Academic Press, London, UK, 2015.
10. Graham, Erin R., and Robert W. Sanders. "Species-specific photosynthetic responses of symbiotic zoanthids to thermal stress and ocean acidification." *Marine Ecology* 37, no. 2 (2016): 442–458.
11. Pischl, Phyllis H., and Nicholas A. Barber. "Plant responses to arbuscular mycorrhizae under elevated temperature and drought." *Journal of Plant Ecology* 10, no. 4 (2017): 692–701.
12. Brener-Raffalli, Kelly, Camille Clerissi, Jeremie Vidal-Dupiol et al. "Thermal regime and host clade, rather than geography, drive Symbiodinium and bacterial assemblages in the scleractinian coral Pocillopora damicornis sensu lato." *Microbiome* 6, no. 1 (2018): 1–13.

13. Guo, Xue, Xishu Zhou, Lauren Hale et al. "Climate warming accelerates temporal scaling of grassland soil microbial biodiversity." *Nature Ecology & Evolution* 3, no. 4 (2019): 612–619.
14. Parts, Kaarin, Leho Tedersoo, Andreas Schindlbacher et al. "Acclimation of fine root systems to soil warming: Comparison of an experimental setup and a natural soil temperature gradient." *Ecosystems* 22, no. 3 (2019): 457–472.
15. Mueller, Gregory M., John Paul Schmit, Sabine M. Hubndorf Leif Ryvarden et al. "Recommended protocols for sampling macrofungi." In *Biodiversity of Fungi: Inventory and Monitoring Methods*, 168–172, Elsevier Academic Press, Amsterdam; Boston, 2004.
16. Feller, G. "Psychrophilic enzymes: from folding to function and biotechnology." *Scientifica (Cairo)* 2013 (2013): 512840.
17. Cavicchioli, R., T. Charlton, H. Ertan, S. Mohd Omar et al. "Biotechnological uses of enzymes from psychrophiles." *Microbial Biotechnology* 4, no. 4 (2011): 449–460.
18. De Maayer, Pieter, Dominique Anderson, Craig Cary, and Don A. Cowan. "Some like it cold: Understanding the survival strategies of psychrophiles." *EMBO Reports* 15, no. 5 (2014): 508–517.
19. Rizvi, Asfa, Bilal Ahmed, Mohammad Saghir Khan et al. "Psychrophilic Bacterial Phosphate-Biofertilizers: A Novel Extremophile for Sustainable Crop Production under Cold Environment." *Microorganisms* 9, no. 12 (2021): 2451.
20. Margesin, Rosa, and Georges Feller. "Biotechnological applications of psychrophiles." *Environmental Technology* 31, no. 8–9 (2010): 835–844.
21. Gerday, C. "Psychrophily and catalysis." *Biology* 2 (2013): 719–741.
22. Jia, Guodong, Zhiguo Rao, Jie Zhang et al. "Tetraether biomarker records from a loess-paleosol sequence in the western Chinese Loess Plateau." *Frontiers in Microbiology* 4 (2013): 199.
23. Bakermans, Corien, Peter W. Bergholz, Debora F. Rodrigues et al. "Genomic and expression analyses of cold-adapted microorganisms." *Polar Microbiology: Life in a Deep Freeze* (2011): 126–155.
24. Yamanaka, Kunitoshi, Li Fang, and Masayori Inouye. "The CspA family in *Escherichia coli*: Multiple gene duplication for stress adaptation." *Molecular Microbiology* 27, no. 2 (1998): 247–255.
25. Graumann, Peter, and Mohamed A. Marahiel. "Some like it cold: Response of microorganisms to cold shock." *Archives of Microbiology* 166, no. 5 (1996): 293–300.
26. Lee, Juho, Ki-Woong Jeong, Bonghwan Jin et al. "Structural and dynamic features of cold-shock proteins of Listeria monocytogenes, a psychrophilic bacterium." *Biochemistry* 52, no. 14 (2013): 2492–2504.
27. Phadtare, Sangita. "Recent developments in bacterial cold-shock response." *Current Issues in Molecular Biology* 6, no. 2 (2004): 125–136.
28. Duru, Ilhan Cem, Anne Ylinen, Sergei Belanov et al. "Transcriptomic time-series analysis of cold-and heat-shock response in psychrotrophic lactic acid bacteria." *BMC Genomics* 22, no. 1 (2021): 1–16.
29. Sung, Min-Sun, Ha-Na Im, and Kyung-Hee Lee. "Molecular cloning and chaperone activity of DnaK from cold-adapted bacteria, KOPRI22215." *Bulletin of the Korean Chemical Society* 32, no. 6 (2011): 1925–1930.
30. Ferrer, Manuel, Heinrich Lünsdorf, Tatyana N. Chernikova et al. "Functional consequences of single: Double ring transitions in chaperonins: Life in the cold." *Molecular Microbiology* 53, no. 1 (2004): 167–182.
31. Strocchi, Massimo, Manuel Ferrer, Kenneth N. Timmis, and Peter N. Golyshin. "Low temperature-induced systems failure in *Escherichia coli*: Insights from rescue by cold-adapted chaperones." *Proteomics* 6, no. 1 (2006): 193–206.

32. Raymond-Bouchard, Isabelle, Jacqueline Goordial, Yevgen Zolotarov et al. "Conserved genomic and amino acid traits of cold adaptation in subzero-growing Arctic permafrost bacteria." *FEMS Microbiology Ecology* 94, no. 4 (2018): fiy023.
33. Hartl, F. Ulrich, and Manajit Hayer-Hartl. "Converging concepts of protein folding in vitro and in vivo." *Nature Structural & Molecular Biology* 16, no. 6 (2009): 574–581.
34. Kandror, Olga, and Alfred L. Goldberg. "Trigger factor is induced upon cold shock and enhances viability of Escherichia coli at low temperatures." *Proceedings of the National Academy of Sciences* 94, no. 10 (1997): 4978–4981.
35. Piette, Florence, Salvino D'Amico, Gabriel Mazzucchelli et al. "Life in the cold: A proteomic study of cold-repressed proteins in the Antarctic bacterium Pseudoalteromonas haloplanktis TAC125." *Applied and Environmental Microbiology* 77, no. 11 (2011): 3881–3883.
36. Gregson, Benjamin H., Gergana Metodieva, Metodi V. Metodiev et al. "Protein expression in the obligate hydrocarbon-degrading psychrophile Oleispira antarctica RB-8 during alkane degradation and cold tolerance." *Environmental Microbiology* 22, no. 5 (2020): 1870–1883.
37. Turchetti, Benedetta, Gianpiero Marconi, Ciro Sannino et al. "DNA methylation changes induced by cold in psychrophilic and psychrotolerant Naganishia yeast species." *Microorganisms* 8, no. 2 (2020): 296.
38. Yusof, Nur Athirah, Shazilah Kamaruddin, Farah Diba Abu Bakar et al. "Structural and functional insights into TRiC chaperonin from a psychrophilic yeast, *Glaciozyma antarctica*." *Cell Stress and Chaperones* 24, no. 2 (2019): 351–368.
39. Liu, Qing, Hong-Can Liu, Yu-Guang Zhou, and Yu-Hua Xin. "Microevolution and adaptive strategy of psychrophilic species *Flavobacterium bomense* sp. nov. isolated from glaciers." *Frontiers in Microbiology* 10 (2019): 1069.
40. Owttrim, George W. "RNA helicases: Diverse roles in prokaryotic response to abiotic stress." *RNA Biology* 10, no. 1 (2013): 96–110.
41. Cartier, Gwendoline, Florence Lorieux, Frédéric Allemand et al. "Cold adaptation in DEAD-box proteins." *Biochemistry* 49, no. 12 (2010): 2636–2646.
42. Phadtare, Sangita, and Konstantin Severinov. "RNA remodeling and gene regulation by cold shock proteins." *RNA Biology* 7, no. 6 (2010): 788–795.
43. Charollais, Julie, Marc Dreyfus, and Isabelle Iost. "CsdA, a cold-shock RNA helicase from Escherichia coli, is involved in the biogenesis of 50S ribosomal subunit." *Nucleic Acids Research* 32, no. 9 (2004): 2751–2759.
44. Rodrigues, Debora F., Natalia Ivanova, Zhili He et al. "Architecture of thermal adaptation in an Exiguobacterium sibiricum strain isolated from 3 million year old permafrost: A genome and transcriptome approach." *BMC Genomics* 9, no. 1 (2008): 1–17.
45. Wecker, Patricia, Christine Klockow, Andreas Ellrott et al. "Transcriptional response of the model planctomycete *Rhodopirellula baltica* SH1T to changing environmental conditions." *BMC Genomics* 10, no. 1 (2009): 1–16.
46. Watanabe, Tomohiro, Hisaya Kojima, and Manabu Fukui. "Draft genome sequence of a psychrotolerant sulfur-oxidizing bacterium, Sulfuricella denitrificans skB26, and proteomic insights into cold adaptation." *Applied and Environmental Microbiology* 78, no. 18 (2012): 6545–6549.
47. Watanabe, Tomohiro, Hisaya Kojima, and Manabu Fukui. "Draft genome sequence of a psychrotolerant sulfur-oxidizing bacterium, *Sulfuricella denitrificans* skB26, and proteomic insights into cold adaptation." *Applied and Environmental Microbiology* 78, no. 18 (2012): 6545–6549.

48. Rong, Jin-Cheng, Yang Liu, Shuang Yu, et al. "Complete genome sequence of *Paenisporosarcina antarctica* CGMCC 1.6503 T, a marine psychrophilic bacterium isolated from Antarctica." *Marine Genomics* 49 (2020): 100690.
49. Okuyama, Hidetoshi, Yoshitake Orikasa, and Takanori Nishida. "Significance of antioxidative functions of eicosapentaenoic and docosahexaenoic acids in marine microorganisms." *Applied and Environmental Microbiology* 74, no. 3 (2008): 570–574.
50. Yoshida, Kiyohito, Mikako Hashimoto, Ryuji Hori et al. "Bacterial long-chain polyunsaturated fatty acids: Their biosynthetic genes, functions, and practical use." *Marine Drugs* 14, no. 5 (2016): 94.
51. Puja, Gupta, and Vakhlu Jyoti. "Culturable bacterial diversity and hydrolytic enzymes from drass, a cold desert in India." *African Journal of Microbiology Research* 9, no. 31 (2015): 1866–1876.
52. Keto-Timonen, Riikka, Nina Hietala, Eveliina Palonen et al. "Cold shock proteins: A minireview with special emphasis on Csp-family of enteropathogenic Yersinia." *Frontiers in Microbiology* 7 (2016): 1151.
53. Mykytczuk, Nadia, Simon J. Foote, Chris R. Omelon et al. "Bacterial growth at– 15 C; molecular insights from the permafrost bacterium *Planococcus halocryophilus* Or1." *The ISME Journal* 7, no. 6 (2013): 1211–1226.
54. Casillo, Angela, Ermenegilda Parrilli, Filomena Sannino et al. "Structure-activity relationship of the exopolysaccharide from a psychrophilic bacterium: A strategy for cryoprotection." *Carbohydrate Polymers* 156 (2017): 364–371.
55. Benforte, Florencia C., Maria A. Colonnella, Martiniano M. Ricardi et al. "Novel role of the LPS core glycosyltransferase WapH for cold adaptation in the Antarctic bacterium *Pseudomonas extremaustralis*." *PLoS One* 13, no. 2 (2018): e0192559.
56. Chintalapati, S., M. D. Kiran, and S. Shivaji. "Role of membrane lipid fatty acids in cold adaptation." *Cellular and Molecular Biology (Noisy-le-Grand, France)* 50, no. 5 (2004): 631–642.
57. Methé, Barbara A., Karen E. Nelson, Jody W. Deming et al. "The psychrophilic lifestyle as revealed by the genome sequence of *Colwellia psychrerythraea* 34H through genomic and proteomic analyses." *Proceedings of the National Academy of Sciences* 102, no. 31 (2005): 10913–10918.
58. Collins, R. Eric, and Jody W. Deming. "An inter-order horizontal gene transfer event enables the catabolism of compatible solutes by *Colwellia psychrerythraea* 34H." *Extremophiles* 17, no. 4 (2013): 601–610.
59. Nunn, Brook L., Krystal V. Slattery, Karen A. Cameron et al. "Proteomics of *Colwellia psychrerythraea* at subzero temperatures–a life with limited movement, flexible membranes and vital DNA repair." *Environmental Microbiology* 17, no. 7 (2015): 2319–2335.
60. Ghobakhlou, Abdollah-Fardin, Anne Johnston, Linda Harris et al. "Microarray transcriptional profiling of Arctic *Mesorhizobium* strain N33 at low temperature provides insights into cold adaption strategies." *BMC Genomics* 16, no. 1 (2015): 1–14.
61. López, Nancy I., M. Julia Pettinari, Pablo I. Nikel, and Beatriz S. Méndez. "Polyhydroxyalkanoates: Much more than biodegradable plastics." *Advances in Applied Microbiology* 93 (2015): 73–106.
62. Kumar, Vijay, Sanjay Kumar, and Dharam Singh. "Microbial polyhydroxyalkanoates from extreme niches: Bioprospection status, opportunities and challenges." *International Journal of Biological Macromolecules* 147 (2020): 1255–1267.

63. Kadouri, Daniel, Edouard Jurkevitch, and Yaacov Okon. "Involvement of the reserve material poly-β-hydroxybutyrate in *Azospirillum brasilense* stress endurance and root colonization." *Applied and Environmental Microbiology* 69, no. 6 (2003): 3244–3250.
64. Białkowska, Aneta, Edyta Majewska, Aleksandra Olczak, and Aleksandra Twarda-Clapa. "Ice binding proteins: Diverse biological roles and applications in different types of industry." *Biomolecules* 10, no. 2 (2020): 274.
65. Kim, Hak Jun, Jun Hyuck Lee, Young Baek Hur et al. "Marine antifreeze proteins: Structure, function, and application to cryopreservation as a potential cryoprotectant." *Marine Drugs* 15, no. 2 (2017): 27.
66. Davies, Peter L. "Ice-binding proteins: A remarkable diversity of structures for stopping and starting ice growth." *Trends in Biochemical Sciences* 39, no. 11 (2014): 548–555.
67. Gilbert, Jack A., Philip J. Hill, Christine ER Dodd, and Johanna Laybourn-Parry. "Demonstration of antifreeze protein activity in Antarctic lake bacteria." *Microbiology* 150, no. 1 (2004): 171–180.
68. Muryoi, Naomi, Mika Sato, Shoji Kaneko et al. "Cloning and expression of afpA, a gene encoding an antifreeze protein from the arctic plant growth-promoting rhizobacterium *Pseudomonas putida* GR12-2." *Journal of Bacteriology* 186, no. 17 (2004): 5661–5671.
69. Hamid, Burhan, Ravinder Singh Rana, Deepak Chauhan et al. "Psychrophilic yeasts and their biotechnological applications-a review." *African Journal of Biotechnology* 13, no. 22 (2014): 2188–2197.
70. Hamdan, Amira. "Psychrophiles: Ecological significance and potential industrial application." *South African Journal of Science* 114, no. 5–6 (2018): 1–6.
71. Kumari, Megha, Srichandan Padhi, Swati Sharma et al. "Biotechnological potential of psychrophilic microorganisms as the source of cold-active enzymes in food processing applications." *3 Biotech* 11, no. 11 (2021): 1–18.
72. Bhatia, Ravi Kant, Saleem Ullah, Mubasher Zahir Hoque "Psychrophiles: a source of cold-adapted enzymes for energy efficient biotechnological industrial processes." *Journal of Environmental Chemical Engineering* 9, no. 1 (2021): 104607.
73. Margesin, Rosa, and Vanya Miteva. "Diversity and ecology of psychrophilic microorganisms." *Research in Microbiology* 162, no. 3 (2011): 346–361.
74. Koo, Hyunmin, Joseph A. Hakim, Casey D. Morrow et al. "Metagenomic analysis of microbial community compositions and cold-responsive stress genes in selected Antarctic lacustrine and soil ecosystems." *Life* 8, no. 3 (2018): 29.
75. Männistö, Minna K., Marja Tiirola, and Max M. Häggblom. "Bacterial communities in Arctic fields of Finnish Lapland are stable but highly pH-dependent." *FEMS Microbiology Ecology* 59, no. 2 (2007): 452–465.
76. Margesin, Rosa, Franz Schinner, Jean-Claude Marx, and Charles Gerday, eds. *Psychrophiles: from Biodiversity to Biotechnology*. Berlin: Springer, 2008.
77. Simon, Carola, Judith Herath, Stephanie Rockstroh, and Rolf Daniel. "Rapid identification of genes encoding DNA polymerases by function-based screening of metagenomic libraries derived from glacial ice." *Applied and Environmental Microbiology* 75, no. 9 (2009): 2964–2968.
78. Shravage, Bhupendra V., Kannayakanahalli M. Dayananda, Milind S. Patole, and Yogesh S. Shouche. "Molecular microbial diversity of a soil sample and detection of ammonia oxidizers from Cape Evans, Mcmurdo Dry Valley, Antarctica." *Microbiological Research* 162, no. 1 (2007): 15–25.

79. Shen, Liang, Yongqin Liu, Michelle A. Allen et al. "Linking genomic and physiological characteristics of psychrophilic Arthrobacter to metagenomic data to explain global environmental distribution." *Microbiome* 9, no. 1 (2021): 1–15.
80. Lopatina, Anna, Sofia Medvedeva, Sergey Shmakov et al. "Metagenomic analysis of bacterial communities of Antarctic surface snow." *Frontiers in Microbiology* 7 (2016): 398.
81. Babalola, Olubukola O., Bronwyn M. Kirby, Marilize Le Roes-Hill et al. "Phylogenetic analysis of actinobacterial populations associated with Antarctic Dry Valley mineral soils." *Environmental Microbiology* 11, no. 3 (2009): 566–576.
82. Morán-Tejeda, Enrique, Juan Ignacio López-Moreno, and Martin Beniston. "The changing roles of temperature and precipitation on snowpack variability in Switzerland as a function of altitude." *Geophysical Research Letters* 40, no. 10 (2013): 2131–2136.
83. Joshi, Divya, Saurabh Kumar, Deep Chandra Suyal, and Reeta Goel. "The microbiome of the Himalayan ecosystem." In *Mining of Microbial Wealth and Metagenomics*, pp. 101–116. Springer, Singapore, 2017.
84. Oh, Han Na, Doyoung Park, Hoon Je Seong et al. "Antarctic tundra soil metagenome as useful natural resources of cold-active lignocelluolytic enzymes." *Journal of Microbiology* 57, no. 10 (2019): 865–873.
85. Adhikari, Priyanka, Rahul Jain, Avinash Sharma, and Anita Pandey. "Plant growth promotion at low temperature by phosphate-solubilizing *Pseudomonas* spp. isolated from high-altitude Himalayan soil." *Microbial Ecology* 82, no. 3 (2021): 677–687.
86. Chen, Weimin, Fan Yang, Lu Zhang, and Junman Wang. "Organic acid secretion and phosphate solubilizing efficiency of *Pseudomonas* sp. PSB12: Effects of phosphorus forms and carbon sources." *Geomicrobiology Journal* 33, no. 10 (2016): 870–877.
87. Vyas, Pratibha, and Arvind Gulati. "Organic acid production in vitro and plant growth promotion in maize under controlled environment by phosphate-solubilizing fluorescent *Pseudomonas*." *BMC Microbiology* 9, no. 1 (2009): 1–15.
88. Gulati, Arvind, Natasha Sharma, Pratibha Vyas et al. "Organic acid production and plant growth promotion as a function of phosphate solubilization by *Acinetobacter rhizosphaerae* strain BIHB 723 isolated from the cold deserts of the trans-Himalayas." *Archives of Microbiology* 192, no. 11 (2010): 975–983.
89. Vyas, Pratibha, Robin Joshi, K. C. Sharma et al. "Cold-adapted and rhizosphere-competent strain of *Rahnella* sp. with broad-spectrum plant growth-promotion potential." *Journal of Microbiology and Biotechnology* 20, no. 12 (2010): 1724–1734.
90. Balcazar, Wilvis, Johnma Rondón, Marcos Rengifo et al. "Bioprospecting glacial ice for plant growth promoting bacteria." *Microbiological Research* 177 (2015): 1–7.
91. Zaheer, Ahmad, Arif Malik, Ahmad Sher et al. "Isolation, characterization, and effect of phosphate-zinc-solubilizing bacterial strains on chickpea (*Cicer arietinum* L.) growth." *Sau Journal of Biological Sciences* 26, no. 5 (2019): 1061–1067.
92. Azaroual, Salah Eddine, Zakaria Hazzoumi, Najib El Mernissi et al. "Role of inorganic phosphate solubilizing bacilli isolated from moroccan phosphate rock mine and rhizosphere soils in wheat (*Triticum aestivum* L) phosphorus uptake." *Current Microbiology* 77, no. 9 (2020): 2391–2404.
93. Rfaki, Abderrazak, Omar Zennouhi, Fatima Zahra Aliyat et al. "Isolation, selection and characterization of root-associated rock phosphate solubilizing bacteria in moroccan wheat (*Triticum aestivum* L.)." *Geomicrobiology Journal* 37, no. 3 (2020): 230–241.
94. Chawngthu, Lalrampani, Remruattluanga Hnamte, and R. Lalfakzuala. "Isolation and characterization of rhizospheric phosphate solubilizing bacteria from wetland paddy field of Mizoram, India." *Geomicrobiology Journal* 37, no. 4 (2020): 366–375.

14 Ecology and remediation of soil
The steps toward modern soil biotechnology

*Rakesh Goswami, Sriya Choudhury, Bidyut Bandyopadhyay, Agniswar Sarkar and Mousumi Saha**

CONTENTS

14.1 INTRODUCTION

Soil microbial communities are mostly diverse, and they impact the maintenance of many ecosystem activities, like the nitrogen cycle, soil structure, primary production, litter decomposition, climate regulation, and so on. The ability of soils to sustain such processes is referred to as 'soil health/quality,' and the soil microbiome has been proposed as a possible biomarker for determining soil quality. Most of the microorganisms rely heavily on the quality of numerous characteristics with which they interact within their ecological niche to survive and thrive. Soil

* Corresponding author e-mail: mou.cute@gmail.com; mousumi.saha@adamasuniversity.ac.in

DOI: 10.1201/9781003354031-14

health is critical in defining a habitat's members, determining their productivity, persistence and longevity. All living forms are affected by the quality of soil in which they live, regardless of their complexity.[1]

As we concentrate our research studies toward one of the most important crises confronting our planet, climate change is affecting soil microbiomes. Most soil microflora have developed evolutionary techniques for dealing with climate changes. SMCs (synthetic microbial consortia) were created to improve plant growth and quality, and they were utilised to improve the soil microbiome for superior crops. When forming a microbial connection, it's important to examine the microorganisms' ability to adapt to their new surroundings. It has been found that soil's active bioformulation can reduce the stress of climate warming by allowing more carbon dioxide to be stored, thereby decreasing the negative effects of climate change. The agricultural production and economy of a country are also dependent on soil components, and their initiator would result in barren, unproductive regions.[2] Various studies have shown that many organisms have received rigorous assessment for use in agricultural production following the innovation and recognition of several beneficial elements provided by a majority of environmental microbes, resulting in various biological methods and products available to farmers. Advances in biotechnology were sold primarily as financial incentives for higher crop yields and more efficient animal growth by supplementing or reducing synthetic chemical inputs (Figure 14.1). As a result, there is a

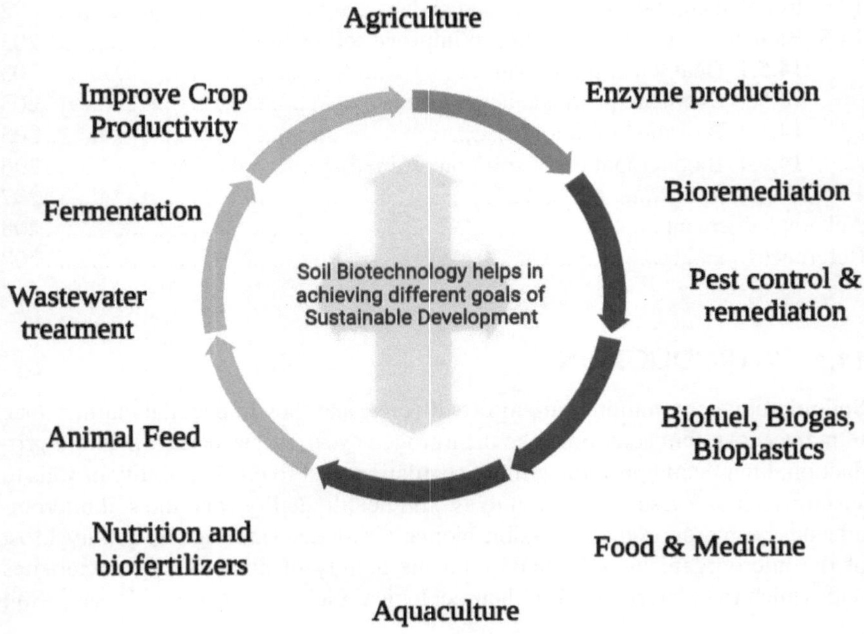

FIGURE 14.1 Representation of different aspects of application of modern soil biotechnology that help in achieving different goals of sustainable development in today's world.

scarcity of data on the direct effects of biotechnology on environmental services and soil health.[3]

Microbes are one of the most important contributors to improving soil health because they degrade contaminants. Microbes' versatility in detoxifying a wide range of contaminants, eco-friendliness, capability in enriching the soil with nutrients, survival in even difficult situations, generation of plant growth-promoting compounds, ease of treatment, and lack of harmful end products are all advantages of using them.[4] Soil health restoration is critical. Microorganisms are well known for their importance in the remediation of soil contaminants such as heavy metals, herbicides and hydrocarbons. However, the native bacteria's power to digest these contaminants is restricted, and the process will be gradual. The discovery of the potential for utilising bioengineered microorganisms in environmental engineering for improving soil health has created major developments in this field. Initiatives combining genetically modified organisms (GMOs), soil additives and plant growth-promoting rhizobacteria will give an effective solution to soil health restoration.[5]

The main objective of the current review is to decipher the possibilities of using biotransformation, microbial detoxification and bioremediation to detox various xenobiotics from the soil while also elevating the soil fertility.

14.2 SOIL MICROBIOLOGY AND HEALTH IMPROVEMENT

Most of the soil microorganisms are active soil engineers. They make fertile soil ready for plant development by ensuring the availability of nutritional factors and producing the growth regulators that plants require most of the time (Figure 14.2).

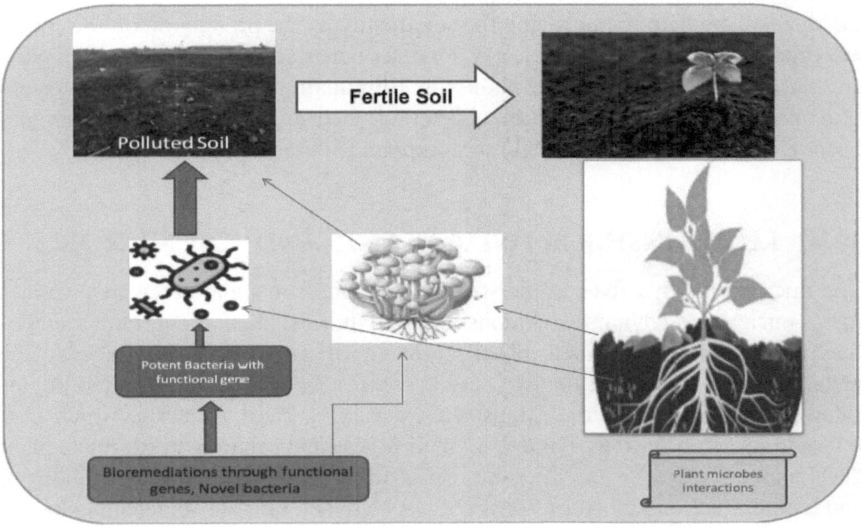

FIGURE. 14.2 Schematic diagram of soil–microbe interaction: representing functional correlation between the soil characteristics and soil microbiome.

The soil's suitability for agriculture and cultivation is improved by the combined activity of plants and microorganisms.[6] Rhizosphere bacteria have been advocated to increase soil fertility in place of chemical fertilisers for achieving sustainable plant development.[7] Several studies have shown that microbes such as *Aspergillus* sp. and *Raoultella ornithinolytica* break down pesticides such as profenofos, endosulfan, pyrethroid, lambda-cypermethrin and deltamethrin.[8] Several studies have reported specific microbes with potential to degrade a wide range of xenobiotics, e.g., plastics, hydrocarbons, surfactants and polychlorinated biphenyls.[5] There is a lot of potential for crop residue, agro-industrial waste and untreated industrial waste (phosphogypsum, basic slag, etc.), which can be used as soil supplements to supply nutrients to crops and maintain healthy soil.[9] Furthermore, innovative techniques for soil health management are prioritised rather than merely relying on manures and fertilisers to provide crop nutrition.[11]

14.3 MICROBIAL XENOBIOTIC DEGRADATION AND PLANT GROWTH-PROMOTING AGENTS

Microbes help in the degradation of pollutants, contaminants and toxic elements and also play a significant role in controlling different xenobiotics' effects on soil and plant health Soil fertility is enhanced by microbial activity. *Pseudomonas* sp. can solubilise phosphate, produce indole acetic acid and fix nitrogen. Thus, *Pseudomonas* sp. helps in improving soil fertility and stimulating plant growth. Rhizobacteria can be used for improvement of soil health, as they remove xenobiotics from roots and also aid in plant growth. They also increase phosphorus and nitrogen availability to plants, ultimately benefiting plant growth and soil health. This method is named *rhizoremediation*. A few bacteria associated with the rhizosphere help in improving the soil quality by the production of hormones, enzymes and plant growth regulators. Phytates act as an antinutrient, reducing the amount of phosphorus available in the soil. Elimination of pollutants and reformation of soil health mediated by rhizobacteria is a cheap, safe and environmentally friendly way to handle the hazardous compounds in the soil.

14.4 RELATIONSHIP BETWEEN PLANTS AND SOIL MICROBES

The microbiome that lives in the soil, rhizosphere, roots and other plant tissues forms intricate and dynamic relationships with the host plant. Plants emit metabolites when stressed to attract microbes that can improve plant resilience. A few instances include the fact that dryness decreases the amount of iron and phytosiderophore available in the rhizosphere, enriching Actinobacteria, which may thrive in low-iron settings, enhancing their fitness and capacity to stimulate plant development.[10] Most of the microbes differ in terms of physiology, metabolism, and temperature and moisture sensitivity. As a result, climate change may have a direct impact on plant microbiome assembly. It has been predicted that host reactions like modifications to plant morphology, physiology, immune response and root exudation affect the rhizosphere microbiome indirectly, whereas the bulk soil microbiome (a reservoir of accessible bacteria that can colonise the plant) is affected directly. Numerous studies have shown that the plant-associated microbiome responds predictably to climate change.[11] Plants have likewise evolved a multilevel microbial management system that allows them to absorb the most beneficial microorganisms into their tissues while also distinguishing between friends and enemies. Interestingly, rhizosphere microbiomes are indirectly affected by changes in plant physiology, morphology, immunological response and root exudation. Climate change consistently regulates the plant–microbiome association. There is growing proof that the development of root length and diameter is constrained by the reciprocal interaction of below-ground photosynthate allocation.[12]

14.5 BIOTECHNOLOGICAL APPROACHES TO IMPROVE SOIL HEALTH

14.5.1 Heavy metal removal

Heavy metals have contaminated five million places around the world, with amounts crossing the legal thresholds.[13] The ecosystem and humans face numerous threats due to heavy metal contamination. Treatment of heavy metals along with microbial remediation can lead to a reduction of industrial usage of chemical bioremediation techniques and the creation of a pollution-free environment in a way that is efficient, environmentally acceptable and economical.[14] Another study found that in order to transport heavy metals from the environment into the cytoplasm of microbes, where they are then further bound by metal-sequestering proteins, bioaccumulation uses a variety of transporters, including ion channels, porins, secondary transporters and active transporters.[15]

Many researchers have already reported that the biosorption method can be used to absorb Cd and Pb metal ions from the environment. Some research has shown that *Chlorella vulgaris* was used as a Cd and Pb absorbent in a biosorption investigation. Biosorption happens whenever monovalent and divalent ions such as Na, Mg and Ca existing on the walls are exchanged for heavy metal ions such as Pb and Cd, which form complexes with functional groups. The cell then transports the Pb and Cd metal ions attached to the cell wall to the deeper cell organelles. Different explanations of organisms with their mechanism of action involved in heavy metal removal are listed in Table 14.1.

14.5.2 Synthetic pesticide degradation

Pesticide use is known to be stressful for the ecosystem, alter soil fertility and perhaps cause toxicity in living beings. Microbial degradation has garnered importance as a prominent intermediate pathway for synthetic pesticides in the environment due to several benefits, such as high elimination effectiveness, environmental friendliness and cost effectiveness. Bioremediation is an efficient

TABLE 14.1
Degradation mechanisms of heavy metals through microbial treatment

Name of organism	Heavy metal degraded	Mechanism of heavy metal removal	Reference
Acinetobacter sp.	Cr	Orthometabolic pathway, sorption of chromium	16
Bacillus cereus	Cr	Bacterial biosorption	17
Bacillus thuringiensis	Cd	Combined remediation, antioxidant activity	18
Staphylococcus sp.	Cr	Adsorption, biosorption	19
Gemella sp.	Pb	Plasmid-mediated heavy metal biotransformation	20
Streptomyces sp.	Pb	Combined remediation, antioxidant activity	21
Turbinaria vulgaris	Cr	Biosorption in acidic conditions	22
Trichoderma sp.	Cd	Spontaneous fungal biosorption	23
Saccharomyces cerevisiae	Cu	Intracellular and surface-based biosorption	24
Pseudomonas sp.	Cu, As, Pb, Co	Bacterial bioflocculation	25
Aspergillus niger	Pb, Cd	Fungal adsorption	26
Enterobacter cloacae	Co	Biosorption by response surface technology	27
Klebsiella pneumoniae	Hg	Biosorption by encapsulated and slime-forming bacteria	28
Cellulosimicrobium sp.	Fe, Zn, Ni, Cu	Reduction of chromium to non-toxic forms and sequestration	29
Chlamydomonas reinhardtii	As	Algal biosorption	30
Acremonium sp.	Al, Mn, Cu	Enzymatic biodegradation	31
Chlorella vulgaris	Cu	Surface-based biosorption	32
Bacillus licheniformis	Fe, Cu, Cd	Extracellular enzymatic biodegradation	33
Bacillus firmus	Pb	Bacterial polysaccharide-mediated adsorption	34
Flavobacterium sp.	Cu	Biosorption by immobilised biomass	35
Rhodobacter capsulatus	Zn	Plasmid-mediated biotransformation	36
Ochrobactrum sp.	Cd	Biosorption by glyphosate-degrading bacteria	37

method for cleaning up soil poisoned by chemical pesticides. Recent studies have focused on discovering the microbiological mechanisms underlying chemical pesticide toxicity, biodegradation and remediation.[38] In field-scale studies, it was discovered that a modified *Escherichia coli* strain carrying atrazine chlorohydrolase was efficient at cleaning up atrazine-polluted soil.[39] From pesticide-polluted soil in Sudan, a few isolated strains were *Bacillus cereus* strain ATCC14579T, *Bacillus safensis* strain FO-36bT and *Bacillus subtilis* subsp. inquosorum strain KCTC13429T. Recently, another study showed that 50 obsolete pesticide storage facilities exist on Kyrgyzstan's territory, storing around 5000 tonnes of these hazardous chemicals. The OPH gene, which codes for organophosphorus hydrolase, breaks down organophosphate pesticides in soil. It is better to use an OPH-engineered *E. coli* for soil decontamination, since it secretes OPH protein into the periplasm and has a 1.8-fold increase in activity.[40]

14.5.3 Biotransformation

Soil pollution and soil quality degradation have become major problems in the modern world (Figure 14.3).

Biotransformation refers to the process of biochemical moderation of the chemical compounds present in a system from one form to another in order to reduce their toxicity and persistence. Plants can be utilised to improve soil health

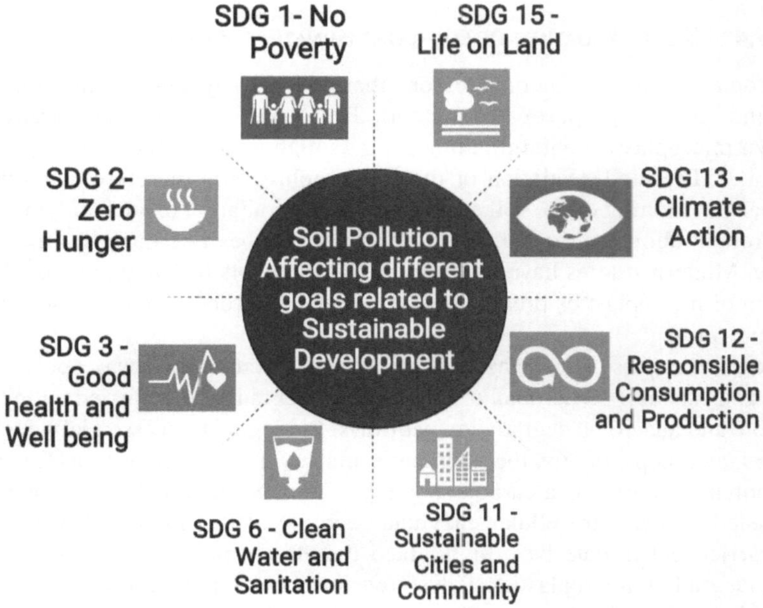

FIGURE 14.3 Depiction of how soil pollution is affecting 8 out of 17 Sustainable Development Goals (SDGs).

by different biotransformation techniques, like phytoextraction, phytodegradation and phytodesalination.[41] Phytoextraction is the process by which toxic chemicals undergo imbibition by the plant roots and are then translocated to those parts of the plant that are metabolically inactive. The toxic metal cations are absorbed by the plant from the soil, forming a metal–ligand complex in the plant cell, followed by the translocation of these complexes to various parts of the plant cell, like the cell wall and vacuole. The best-suited plant species that have been utilised for phytoextraction generally have fast growth and produce high amounts of biomass.[42] Some of the best examples of plants that have phytoextraction capability include *Pelargonium hortorum* for Pb,[43] *Nicotiana tabacum* for Cd,[44] *Zea mays* for Ti,[45] etc.

Saline soil has a considerable negative impact on plant physiology and can cause dehydration and saline stress in the plant body. Phytodesalination is the process by which certain salt-tolerant plant species are capable of extracting salt from the saline soil to reduce its salinity and render the soil productive again.[46] Halophytic plants are most commonly chosen for desalination purposes. They can take up sodium and chloride ions from the saline soil and accumulate them in their roots. Another mechanism by which they can remediate the saline soil is by selectively leaching the salts from the saline soil using their expansive root system. Some common species used for desalination purposes are *Ipomoea aquatica*, *Alternanthera philoxeroides* and *Ludwigia adscendens* for removing sodium ions[47] and *Puccinellia nuttalliana* and *Typha latifolia* for removing chloride ions.[48]

14.5.4 Biodegradation of Soil-Contaminating Microplastics

Microplastics act as major stressors that can heavily impact the biochemical and physical properties of the soil. Thoroughly contaminated areas have shown microplastic contamination levels as high as 7% of the total dry weight of soil.[49] The biodegradation of these microplastics is therefore essential for proper functioning of the soil ecosystem. Biodegradation of such polymers can be brought about by the action of different microbes like bacteria, fungi and algae. Modern studies have highlighted the protocols for enzymatic biodegradation of microplastics present in soil samples in order to improve soil health (Figure 14.4).

Some examples of such enzymes include lipase, cutinase, laccase, polyurethanase and esterase.[28] Microbial microplastic biodegradation is carried out by both aerobic and anaerobic degradation pathways. Biodegradation essentially includes three major steps. Firstly, the microbe is attached to the polymer surface. Then, the polymer is used as a carbon source by the microbe. Finally, the polymer is degraded by the extracellular enzymatic activity of the microbe.[29] Many studies carried out to date have highlighted the effectiveness of various microbes in biodegrading microplastics. Polystyrene has been reported to be significantly degraded by *Pseudomonas* sp., *Curvularia* sp. and *Exiguobacterium* sp.[30] *Bacillus cereus* and *Sporosarcina globispora* have demonstrated high depolymerisation

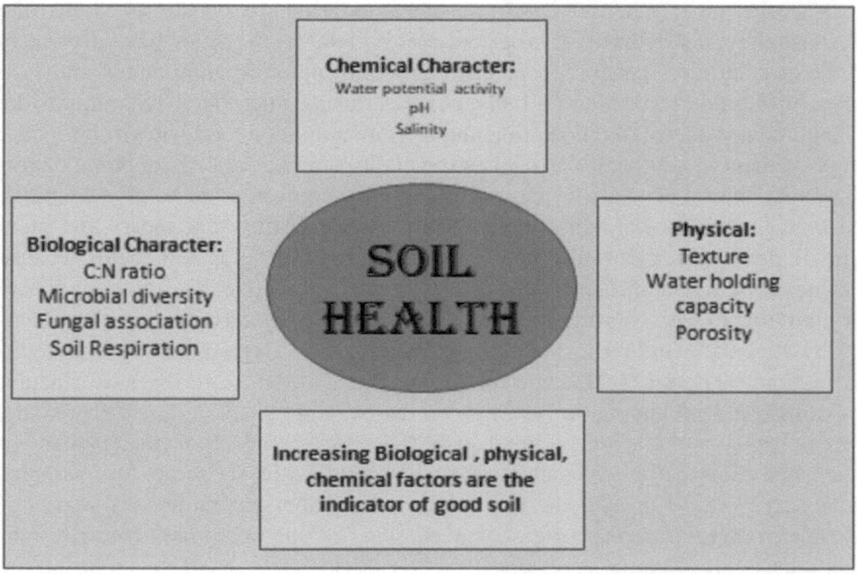

FIGURE 14.4 Illustrative presentation of various physicochemical and biological parameters of soil health.

activity against polypropylene.[31] *Pseudomonas citronellolis* and *Bacillus flexus* have shown a notable degradation efficiency of 19% within 30 days of their incubation.[32]

14.6 CONCLUSION

Soil biotechnology has garnered tremendous importance and relevance in modern times. Advances in soil biotechnology have led to the development of several bioengineering methods to improve the quality of soil. These improvements have on many occasions reflected direct as well as indirect influence on other major aspects of our environment, like sustainable agriculture, wastewater treatment, reduction in soil pollution, management of plant pathogens and improvement in the quality of drinking water. Genetic engineering and molecular techniques for soil health analysis and improvement certainly have an edge over the traditional methods because they are faster, more effective and give better outputs. This chapter has mainly dealt with the relation between soil microbiology, plant development and soil health, followed by the exploration of various biotechnological methods for soil quality improvement, like microbial xenobiotic degradation, introduction of plant growth-promoting factors, bioremediation, synthetic pesticide degradation, biotransformation and biodegradation of microplastics that contaminate soil.

Microbes have been used in various bioengineering approaches for a long time because they can detoxify a variety of soil pollutants. Qualities like survival in difficult situations, generation of plant growth-promoting compounds and lack of harmful end products are all advantages of using microbes.[4] Techniques like bioaugmentation, rhizoremediation and phytoremediation have shown significant improvement in soil health as well as the plants growing in it. Xenobiotic degradation by microbes leads to the production of several intermediates, which can be further studied for their beneficial role as soil fertility enhancers and plant growth promoters. Plants also play a major part in influencing soil health because of their wide range of interactions (competitive, exploitative, neutral, commensal or mutualistic) with soil microbiome. Plants have been seen to improve soil quality by various techniques, like phytoextraction, phytodegradation and phytodesalination.[19] There is still scope for improvement in boosting the biosurfactant capability and enhancing the heavy metal removal efficiency in this field of study. More clarity can be achieved in terms of the different molecular mechanisms of microbial biosorption by conducting further studies on the same. Soil samples collected from all around the world have shown that bioengineered strains of *Escherichia coli*, *Bacillus safensis* and *Bacillus cereus* were able to destroy synthetic pesticides to quite an extent.[33] Some current research is also exploring the capability of microbes in the breakdown of microplastics in soil, like polystyrene, polypropylene and polyethylene terephthalate (PET), and their conversion into non-toxic by-products. However, further upgradation can be achieved in terms of the period of incubation and the degradation efficiency. Thus, on detailed analysis of the already prevalent biotechnological methods for improving soil health, we have come to the conclusion that though significant improvement has been achieved in this field in the past few years, there still exists a lacuna, especially in understanding the exact molecular mechanisms of these techniques and enhancement in their efficiency and outputs. There is also scope for improvement in terms of conversion of the neutral by-products of biotransformation and bioremediation into substances that are useful in improving soil fertility and promoting plant growth. Thus, further studies on the aforementioned topics are advised, as they will be beneficial to society and at the same time give rise to a better understanding of soil biotechnology.

ACKNOWLEDGEMENT

The authors wish to thank Oriental Institute of Science and Technology for providing academic support to prepare the manuscript. The authors also convey their thanks to Dr. Anindya Sundar Panja for proofreading the manuscript.

REFERENCES

1. Sandipan, Samaddar, Karp Daniel, Schmidt Radomir et al. “Role of soil in the regulation of human and plant pathogens: Soils’ contributions to people.” *Philosophical Transactions of the Royal Society B* 376 no. 1834 (2021): 20200179.

2. Kremer, Robert J. "Biotechnology impacts on soil and environmental services." In *Soil Health and Intensification of Agroecosytems*, pp. 353–375. Academic Press, 2017.
3. Bunea, Anca, I. Pacurar, and Vasile Dulf. "Study concerning the soil resources used in agriculture and forestry worldwide." *ProEnvironment* 13 (2020): 39–43.
4. Mishra, Sandhya, Ziqiu Lin, Shimei Pang et al. "Recent advanced technologies for the characterization of xenobiotic-degrading microorganisms and microbial communities." *Frontiers in Bioengineering and Biotechnology* 9 (2021): 632059.
5. Cf, Sini Francis, Sharrel Rebello, Embalil Mathachan et al. "Bioprospecting of gut microflora for plastic biodegradation." *Bioengineered* 12, no. 1 (2021): 1040–1053.
6. Lyu, Dongmei, Levini A. Msimbira, Mahtab Nazari et al. "The coevolution of plants and microbes underpins sustainable agriculture." *Microorganisms* 9, no. 5 (2021): 1036.
7. Nehal, Nikita, Utkarsh Singh Rathore, and Nitish Sharma. "Microbes and soil health for sustainable crop production." In *Microbial Metatranscriptomics Belowground*, pp. 581–613. Springer, Singapore, 2021.
8. Zhang, Xiaoqing, Xiangxiang Hao, Shanshan Huo et al. "Isolation and identification of the *Raoultella ornithinolytica*-ZK4 degrading pyrethroid pesticides within soil sediment from an abandoned pesticide plant." *Archives of Microbiology* 201, no. 9 (2019): 1207–1217.
9. Shahane, Amit Anil, and Yashbir Singh Shivay. "Soil health and its improvement through novel agronomic and innovative approaches." *Frontiers in Agronomy* 3 (2021): 680456.
10. Xu, Ling, Zhaobin Dong, Dawn Chiniquy et al. "Genome-resolved metagenomics reveals role of iron metabolism in drought-induced rhizosphere microbiome dynamics." *Nature Communications* 12, no. 1 (2021): 1–17.
11. Wipf, Heidi M.-L., Thảo-Nguyên Bùi, and Devin Coleman-Derr. "Distinguishing between the impacts of heat and drought stress on the root microbiome of Sorghum bicolor." *Phytobiomes Journal* 5, no. 2 (2021): 166–176.
12. Qiu, Yunpeng, Lijin Guo, Xinyu Xu, Lin Zhang, Kangcheng Zhang, Mengfei Chen, Yexin Zhao et al. "Warming and elevated ozone induce tradeoffs between fine roots and mycorrhizal fungi and stimulate organic carbon decomposition." *Science Advances* 7, no. 28 (2021): eabe9256.
13. Pal, Amit Kumar, and Chandan Sengupta. "Isolation of cadmium and lead tolerant plant growth promoting rhizobacteria: *Lysinibacillus varians* and *Pseudomonas putida* from Indian Agricultural Soil." *Soil and Sediment Contamination: An International Journal* 28, no. 7 (2019): 601–629.
14. Verma, Samakshi, and Arindam Kuila. "Bioremediation of heavy metals by microbial process." *Environmental Technology & Innovation* 14 (2019): 100369.
15. Diep, Patrick, Radhakrishnan Mahadevan, and Alexander F. Yakunin. "Heavy metal removal by bioaccumulation using genetically engineered microorganisms." *Frontiers in Bioengineering and Biotechnology* 6 (2018): 157.
16. Bhattacharya, Amrik, Anshu Gupta, Amarjeet Kaur, and Darshan Malik. "Efficacy of *Acinetobacter* sp. B9 for simultaneous removal of phenol and hexavalent chromium from co-contaminated system." *Applied Microbiology and Biotechnology* 98, no. 23 (2014): 9829–9841.
17. Singh, Neha, Tuhina Verma, and Rajeeva Gaur. "Detoxification of hexavalent chromium by an indigenous facultative anaerobic Bacillus cereus strain isolated from tannery effluent." *African Journal of Biotechnology* 12, no. 10 (2013).
18. Jiang, Juan, Hongying Liu, Qiao Li et al. "Combined remediation of Cd–phenanthrene co-contaminated soil by *Pleurotus cornucopiae* and *Bacillus thuringiensis*

FQ1 and the antioxidant responses in *Pleurotus cornucopiae*." *Ecotoxicology and Environmental Safety* 120 (2015): 386–393.
19. Kumar, Rajender, Divya Bhatia, Rajesh Singh et al. "Sorption of heavy metals from electroplating effluent using immobilized biomass *Trichoderma viride* in a continuous packed-bed column." *International Biodeterioration & Biodegradation* 65, no. 8 (2011): 1133–1139.
20. Marzan, Lolo Wal, Mehjabeen Hossain, Sohana Akter Mina et al. "Isolation and biochemical characterization of heavy-metal resistant bacteria from tannery effluent in Chittagong city, Bangladesh: Bioremediation viewpoint." *The Egyptian Journal of Aquatic Research* 43, no. 1 (2017): 65–74.
21. Boddu, Sumalatha, Venkata Narayana Alugunulla, John Babu Dulla et al. "Estimation of biosorption characteristics of chromium (VI) from aqueous and real tannery effluents by treated T. vulgaris: Experimental assessment and statistical modelling." *International Journal of Environmental Analytical Chemistry* (2020): 1–20.
22. Bazrafshan, Edris, Amin Allah Zarei, and Ferdos Kord Mostafapour. "Biosorption of cadmium from aqueous solutions by Trichoderma fungus: Kinetic, thermodynamic, and equilibrium study." *Desalination and Water Treatment* 57, no. 31 (2016): 14598–14608.
23. Amirnia, Shahram, Madhumita B. Ray, and Argyrios Margaritis. "Heavy metals removal from aqueous solutions using Saccharomyces cerevisiae in a novel continuous bioreactor–biosorption system." *Chemical Engineering Journal* 264 (2015): 863–872.
24. Azzam, Ahmed M., and Ahmed Tawfik. "Removal of heavy metals using bacterial bio-flocculants of *Bacillus* sp. and *Pseudomonas* sp." *Journal of Environmental Engineering and Landscape Management* 23, no. 4 (2015): 288–294.
25. Iram, Shazia, Rabia Shabbir, Hunnia Zafar, and Mehwish Javaid. "Biosorption and bioaccumulation of copper and lead by heavy metal-resistant fungal isolates." *Arabian Journal for Science and Engineering* 40, no. 7 (2015): 1867–1873.
26. Jafari, Seyed Ali, Sama Cheraghi, Maryam Mirbakhsh et al. "Employing response surface methodology for optimization of mercury bioremediation by *Vibrio parahaemolyticus* PG02 in coastal sediments of Bushehr, Iran." *CLEAN–Soil, Air, Water* 43, no. 1 (2015): 118–126.
27. Al-Garni, Saleh M., Khaled M. Ghanem, and Ahmed S. Ibrahim. "Biosorption of mercury by capsulated and slime layerforming Gram-ve bacilli from an aqueous solution." *African Journal of Biotechnology* 9, no. 38 (2010): 6413–6421.
28. Bhati, Tushar, Rahul Gupta, Nisha Yadav et al. "Assessment of bioremediation potential of *Cellulosimicrobium* sp. for treatment of multiple heavy metals." *Microbiology and Biotechnology Letters* 47, no. 2 (2019): 269–277.
29. Saavedra, Ricardo, Raúl Muñoz, María Elisa Taboada et al. "Comparative uptake study of arsenic, boron, copper, manganese and zinc from water by different green microalgae." *Bioresource Technology* 263 (2018): 49–57.
30. Ma, Xiao-kui, and Hala Fam. "Heavy metal ions affecting the removal of polycyclic aromatic hydrocarbons by fungi with heavy-metal resistance." *Applied Microbiology and Biotechnology* 98, no. 23 (2014): 9817–9827.
31. Goher, Mohamed E., Abd El-Monem A.M., Amaal M. Abdel-Satar et al. "Biosorption of some toxic metals from aqueous solution using non-living algal cells of *Chlorella vulgaris*." *Journal of Elementology* 21, no. 3 (2016).
32. Samarth, Deepti Pranay, Chandrashekhar Jangluji Chandekar, and Rama Kaustubh Bhadekar. "Biosorption of heavy metals from aqueous solution using *Bacillus*

licheniformis." *International Journal of Pure and Applied Sciences and Technology* 10, no. 2 (2012): 12.
33. Salehizadeh, H., and S. A. Shojaosadati. "Removal of metal ions from aqueous solution by polysaccharide produced from *Bacillus firmus*." *Water Research* 37, no. 17 (2003): 4231–4235.
34. Magnin, Jean-Pierre, Nicolas Gondrexon, and John C. Willison. "Zinc biosorption by the purple non-sulfur bacterium *Rhodobacter capsulatus*." *Canadian Journal of Microbiology* 60, no. 12 (2014): 829–837.
35. Khadivinia, Elnaz, Hakimeh Sharafi, Faranak Hadi et al. "Cadmium biosorption by a glyphosate-degrading bacterium, a novel biosorbent isolated from pesticide-contaminated agricultural soils." *Journal of Industrial and Engineering Chemistry* 20, no. 6 (2014): 4304–4310.
36. Huang, Xing, and Guining Lu. "Bioremediation of chemical pesticides polluted soil." *Frontiers in Microbiology* 12 (2021): 682343.
37. Strong, Lisa C., Hugh McTavish, Michael J. Sadowsky, and Lawrence P. Wackett. "Field-scale remediation of atrazine-contaminated soil using recombinant Escherichia coli expressing atrazine chlorohydrolase." *Environmental Microbiology* 2, no. 1 (2000): 91–98.
38. Kang, Dong Gyun, Suk Soon Choi, and Hyung Joon Cha. "Enhanced biodegradation of toxic organophosphate compounds using recombinant *Escherichia coli* with sec pathway-driven periplasmic secretion of organophosphorus hydrolase." *Biotechnology Progress* 22, no. 2 (2006): 406–410.
39. Kafle, Arjun, Anil Timilsina, Asmita Gautam et al. "Phytoremediation: Mechanisms, plant selection and enhancement by natural and synthetic agents." *Environmental Advances* (2022): 100203.
40. Bian, Fangyuan, Zheke Zhong, Xiaoping Zhang et al. "Bamboo: An untapped plant resource for the phytoremediation of heavy metal contaminated soils." *Chemosphere* 246 (2020): 125750.
41. Manzoor, Maria, Iram Gul, Iftikhar Ahmed et al. "Metal tolerant bacteria enhanced phytoextraction of lead by two accumulator ornamental species." *Chemosphere* 227 (2019): 561–569.
42. Yang, Yang, Yichen Ge, Pengfei Tu et al. "Phytoextraction of Cd from a contaminated soil by tobacco and safe use of its metal-enriched biomass." *Journal of Hazardous Materials* 363 (2019): 385–393.
43. Huang, Xing, and Guining Lu. "Bioremediation of chemical pesticides polluted soil." *Frontiers in Microbiology* 12 (2021): 682343.
44. Arif, Yamshi, Priyanka Singh, Husna Siddiqui et al. "Salinity induced physiological and biochemical changes in plants: An omic approach towards salt stress tolerance." *Plant Physiology and Biochemistry* 156 (2020): 64–77.
45. Islam, Md S., M. Md L. Hosen, and Md N. Uddin. "Phytodesalination of saline water using *Ipomoea aquatica, Alternanthera philoxeroides* and *Ludwigia adscendens*." *International Journal of Environmental Science and Technology* 16, no. 2 (2019): 965–972.
46. Xu, Qian, Sylvie Renault, and Qiuyan Yuan. "Phytodesalination of landfill leachate using *Puccinellia nuttalliana* and *Typha latifolia*." *International Journal of Phytoremediation* 21, no. 9 (2019): 831–839.
47. Fuller, Stephen, and Anil Gautam. "A procedure for measuring microplastics using pressurized fluid extraction." *Environmental Science & Technology* 50, no. 11 (2016): 5774–5780.

48. Kliem, Silvia, Marc Kreutzbruck, and Christian Bonten. "Review on the biological degradation of polymers in various environments." *Materials* 13, no. 20 (2020): 4586.
49. Elahi, Amina, Dilara Abbas Bukhari, Saba Shamim, and Abdul Rehman. "Plastics degradation by microbes: A sustainable approach." *Journal of King Saud University-Science* 33, no. 6 (2021): 101538.

15 Machine learning to explore the psychrophilic soil microbiome

Shakira Ghazanfar, Rameesha Abid, Abeer Khan, Muhammad Yasir Akbar, Iqra Asif, Reem Nadeem, Wajya Ajmal and Adnan Ansari*

CONTENTS

15.1 INTRODUCTION

The biogeochemical cycle of micronutrients and macronutrients, including several other elements that are fundamental for the growth of animal and plant life, is controlled by the soil microbiome.[1] The structure, physiological activities and composition of soil microbiomes are greatly affected by abiotic stresses along with low temperatures. Psychrophilic bacteria that can ameliorate low-temperature conditions while maintaining physiological activities are often found in severe cold-temperature habitats.[2]

* Corresponding author e-mail: shakira_akmal@yahoo.com

DOI: 10.1201/9781003354031-15

All the constantly cold environments, including mountains, deep sea and polar regions, are completely colonized by psychrophilic microorganisms. Based on the optimal growth temperature, these organisms are classified as psychrotrophic or psychrotolerant.[3,4] However, we agree that depending on the microorganism, there is a continuum in temperature adaptability based on extreme or common growth temperature variations, and to distinguish every microorganism that shows reasonable growth close to the freezing point of water, the common term *psychrophiles* is assigned in this chapter. Because of these significant characteristics, psychrophiles have been able to overcome two difficulties: (1) the low temperature, as even a slightly low temperature causes a great change in the rate of biochemical reactions, and (2) the viscosity of water in the environment, which increases more than twofold in the range from 37 °C to 0 °C. A psychropiezophilic organism known as *Moritella profunda* has shown an adaptation to living in a cold deep sea, as it exhibits the maximum rate of growth temperature at 12 °C and maximal rate of growth at 2 °C; hence, it marks a noteworthy example of adaptation.[3,5] Therefore, it can be deduced that some supramolecular structures or enzymes exhibit a changed conformation, which causes adverse effects on metabolic flux at the low temperature of 2 °C.[3]

Explaining the relation between certain functions of the ecosystem and microbes is still challenging in ecology due to several noticeable macroscopic qualities governed by the microbiome. Deriving the metagenomic data, which represents the taxonomic profile of fungi and bacteria in communities of microbes, has become convenient because of advancements in DNA sequencing in recent years. As a result, this has provided an opportunity to understand and determine the constituents of the microbiome that cause alterations in macroscopic quantities.

A source for multidimensional data analysis is provided by machine learning methods,[6,7] and these methods also help to deduce the relation between microbial taxa, including other metagenomic qualities like metabolic pathways or gene families and other environmental characteristics. Moreover, the random forest model is considered to be an efficacious method of machine learning, as an increased level of precision is exhibited, having a wide range of 16S rRNA datasets for the classification of body habitat, disease conditions and host.[8] To distinguish the subtype of microbial taxa which has significant profusions that offer predictions of selected continual variables, neural network models and random forest models are also utilized, as commonly used software packages of machine learning target the binary identification of microbial datasets. The characteristic importance can be assessed through the amalgamation of neural networks and the random forest.[9]

This chapter describes in detail the predictive models used for the extraction of psychrophilic markers based on machine learning. Previous studies have employed complete proteomes of psychrophiles or filtered protein sequences established on certain procedures for comparative analysis, as the accessibility of psychrophilic protein is still smaller in number in comparison to the existing sequences.

15.2 MACHINE LEARNING AND SOIL MICROBIAL COMMUNITY DATA ANALYSIS

Machine learning in the context of soil microbial ecology entails developing and analyzing models that make use of functions capable of identifying, categorizing and forecasting outcomes from data. There are different types of machine learning methods, including supervised, unsupervised, semi-supervised and reinforced machine learning.[10] For instance, the objective of supervised machine learning applied to soil microbial community data is frequently to build a decision rule from a set of collected observations in order to predict the outcome of input data. In microbiome investigations, the input data and their assigned values take the form of a frequency count matrix of detected microbial species from a sample. It is common to refer to samples as observations and input variables as features. As discussed, there are several machine learning methods, but supervised and unsupervised approaches are used for microbial datasets. The key difference between supervised and unsupervised machine learning is that the supervised method finds a best-fit decision boundary between features and response variables, whereas unsupervised methods separate samples using features without reference to response labels and predict to which cluster a response may belong.[11] Since each sample in the dataset is given a response label in supervised learning, as opposed to unsupervised approaches, supervised learning is a more elaborate way of investigating marker-gene datasets.

15.2.1 MACHINE LEARNING IN MICROBIOME RESEARCH

Machine learning is a powerful technique that is integrated continuously into microbiome research for insight and detailed understanding of the microbiome. There are multiple applications of machine learning in microbiome research, as depicted in Figure 15.1. One application is identification of microbial taxa within

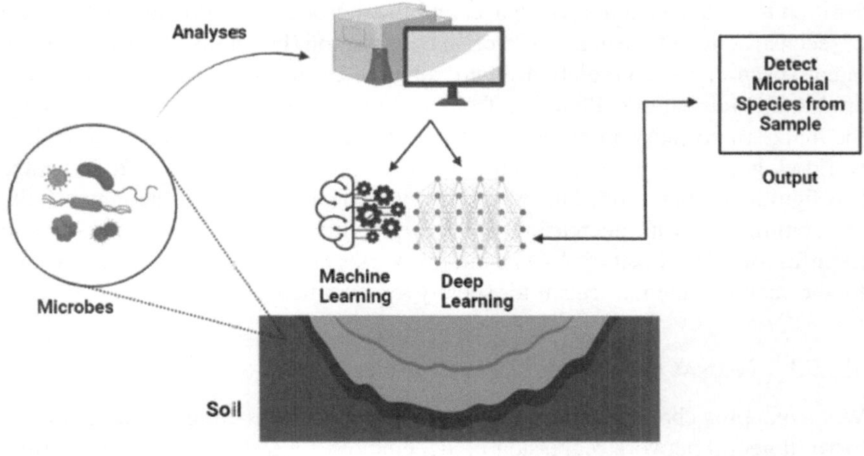

FIGURE 15.1 Application of machine learning in microbial research.

communities impacting the host's phenotypes.[9] To integrate the abundance of metagenomic species for phenotypic prediction, machine learning models like random forest (RF) or support vector machine (SVM) are used; however, productivity prediction is carried out based on soil microbiome composition. Also, machine learning in the soil is used to forecast crop production from soil using numerous soil microbiome data.[12] Various researchers have highlighted the use of machine learning techniques to study the microbiome and metagenomics,[13] as shown in Table 15.1. This chapter separates the two main categories of research. The first comprises studies focused on a single issue, such as the use of microbial composition to forecast soil productivity, contaminants and geochemical features in wells and the presence or absence of disease because of variations in microbial abundance over time, or biomarkers of cancer derived from the human blood microbiome. The second covers global techniques, in which the same tool may be used to analyze various datasets with various forecasting aims.

15.2.2 Deep learning in bioinformatics and microbiome/metagenomics

Deep learning techniques for biological problems have recently emerged.[14] In domains like medical image-based diagnostics and predictions based on genomic sequence analysis, deep learning enhances performance when the input data are pictures or sequences (i.e., deep learning favourable data formats).[15] Improvements in data representation or automatic feature extraction are other areas where deep learning plays a role, such as when features are taken out of electronic health records to determine a patient's risk of illness.[16] Deep learning has not yet been utilized as frequently in microbiome investigations as it has in other bioinformatics issues. According to one study,[17] the most typical job accomplished by machine learning or the deep learning realm is phenotypic prediction from metagenomics data: the development of a novel model that combines information about environmental and mapping features with knowledge about microbial composition to create a deep latent space, and the thorough selection and assessment of reconstruction of the entire microbial composition (hundreds of taxa) from that latent space and the prediction of microbiome composition starting from only environmental features. Real data from a soil microbiome investigation, namely, the maize rhizosphere microbiome, is used to support each of these claims. In a world with an expected population of ten billion people by the year 2050, maize is a significant food crop. This will necessitate tripling food production on finite agricultural soil with decreasing water availability and avoiding the costly and harmful use of fertilisers.[18] As a result, the work's primary area of interest is agri-food research, which is crucial for society's economic well-being.

15.2.3 Neural network regression model

When mapping characteristics in a sample to a continuous target variable, a feed-forward neural network regression model employs several parameterized activation functions arranged in layers. A feed-forward neural network's layer is made

TABLE 15.1
Computational tools used for the detection of psychrophilic microbes

Software	Short Description	Source	Uses
Microbiome Specific Machine Learning Platforms			
MetAML	Metagenomic prediction Analysis based on Machine Learning	Python: https://github.com/segatalab/metaml	DI, FS, ML, IM
SIAMCAT	Statistical Inference of Associations between Microbial Communities and host phenotypes	R package 'SIAMCAT': https://siamcat.embl.de/	DI, FS, ML, IM
mAML	An automated machine learning pipeline with a microbiome repository for human disease classification	Python: https://github.com/yangfenglong/mAML1.0Web:http://lab.malab.cn/soft/mAML	DI, FS, ML, IM
DeepMicro	Deep representation learning for disease prediction based on microbiome data	Python: https://github.com/minoh0201/DeepMicro	DR, ML
MicrobiomeAnalystR	Comprehensive statistical, functional and meta-analysis of microbiome data	R package 'MicrobiomeAnalystR': https://www.microbiomeanalyst.ca/	DI, FS, DR, ML, IM
QIIME2	Quantitative Insights Into Microbial Ecology	https://qiime2.org/	DI, FS, DR, ML,IM
mothur	Microbial community analysis pipeline	http://mothur.org/	DI, FS, DR, ML, IM
General Machine Learning Platforms			
Scikit-learn	Python library for predictive machine learning	Python: https://scikitlearn.org/stable/	DI, FS, DR, ML,IM
Keras	Deep learning library	R package 'keras' Python: https://pypi.org/project/Keras/	DI, FS, DR, ML,IM
caret	Classification And Regression Training	R package 'caret'	DI, FS, DR, ML,IM
mlr	Machine learning in R	R package 'mlr30': https://mlr3.mlrorg.com/	DI, FS, DR, ML,IM
Dtreeviz	Decision tree visualisation	Python: https://github.com/parrt/dtreeviz	FS, ML, IM
ranger	Random forest generator	R package 'ranger'	FS, ML, IM

DI: data visualisation; DR: dimensionality reduction; FS: feature selection; IM: interpretation measure; ML: machine learning up of many nodes that use a nonlinear transformation to the bias parameter plus the product of the inputs from the preceding layer's weight and bias parameters. A learning rate hyper-parameter may be used to control the rate of parameter change during training, and a regularization hyper-parameter can be used to control the cost function to prevent disproportionate parameter values.[19] Using THEANO[20] and Python, a feed-forward neural network regression model is created. Scikit-learn random search technique was used to find the model's hyper-parameters. The model by default consists of a single output layer with a linear activation function and a single hidden layer with 15 nodes that have a sigmoid activation function. The training dataset is used in a randomized hyper-parameter search to determine the ideal hidden layer size, learning rate and regularization coefficient for usage in the training and validation studies. The model uses the mean squared error between the predicted and true values as a cost function. The process is referred to as iteratively training the neural network model, as training data is partitioned into mini-batches and trained over them in each training session. Training comes to an end when the cost function applied to the validation dataset does not decrease after a default of ten training epochs. Results from training and testing are obtained using Pearson's correlation coefficient.

15.2.4 Neural network feature selection

The relevance of features using the neural network model is assessed frequently by weights given to individual features during model training.[21] The gradient of the model output concerning the weights assigned to each feature is utilized to calculate the feature significance vector in the suggested feature selection method. The magnitude of each member shows how important the feature is for predicting the target variable, and the sign indicates whether the feature has a positive or negative influence on the anticipated variable. Each element of the feature significance vectors a single feature. We may refer to the M × J matrix of weights linking each feature to each node in a feed-forward neural network model with M input features as inputs that connect J nodes in the first hidden layer as $\boldsymbol{\theta}^{\mathbf{ln}} \in R^{M \times J}$, where $\boldsymbol{\theta}^{\mathbf{ln}}$ is a subset of the whole parameter set $\boldsymbol{\theta}$. The M × J feature significance matrix $F(\boldsymbol{\theta},x) \in R^{M \times J}$ is provided by the gradient of the model output with respect to $\boldsymbol{\theta}^{\mathbf{ln}}$ and is defined as

$$F_{mj}\left(\theta,x\right)=\frac{\partial}{\partial\theta_{mj}^{ln}}M\left(\theta,x\right)$$

The feature importance vector f (θ ,x) will be referred to as an M-dimensional vector that is created by marginalizing the feature importance matrix across all of the nodes in the first hidden layer. After the model has been trained, the average value of the feature significance vector over the set of training data with K samples to establish the sensitivity of the model to each feature is indicated as the M-dimensional vector $s \in R^{M}$. The bootstrap approach, which randomly chooses 80% of the training dataset over a default of 50 iterations, is used to estimate feature importance in order to increase confidence in the importance ascribed to features. The collection of rated features that may be more confidently ranked has the average feature ranking values across all iterations.

15.2.5 Random forest regression model

By dividing the group of likely selected variables established on the standards of distinct qualities, decision tree-based machine learning methods plot qualities to selected variables.[7] A division in the group of plausible selected target variables is determined by the value of the feature at a point of an internal node along with the node known as a leaf node, which follows an internal node. A group of decision trees formed of an arbitrarily chosen subgroup of feature space is based on the RF method, which calculates the model standard output by limiting the predictions from distinct decision trees. This model can be determined with the mean squared cost function by applying an RF regressor through Scikit-learn.[22] An internal node is divided by two samples, while to be the default at the leaf node, one sample is necessary. A model requires some hyper-parameters, such as 1) the total number of samples necessary to divide an internal node, 2) the total number of samples necessary for a leaf node and 3) the total number of significant features to use in the decision tree. By employing Scikit-learn's randomised search algorithm, hyper-parameters of the model can be adjusted with the training set. In the course of training, the model connects to a community of thousands of decision trees that are trained on arbitrarily targeted sub-samples of the data set. Testing data and identical training can be used from the RF results, which are used to compare neural network models.

15.2.6 Random forest feature selection

An array of values of significant features, similar to the array of feedback features, is yielded by an RF regressor, which is accessible through Scikit-learn.[22] Several decision tree algorithms measure the feature significance by investigating the ability of a feature (frequently denoted as a variable) to divide the possible output labels. To put it simply, an indication of feature significance in a given sample is its potential to reduce possible labels. In addition to this, the significance of the feature is recognised to be related to the bootstrap method employed by arranging random decision trees, keeping the significance of the feature higher for variables whose outcome presents higher prediction performance when included in the decision trees.[23] Feature ranking is characterized by employing a bootstrap method that arbitrarily samples 80% of the training data arranged on a default of 50 repetitions. The highly confident marked features are represented by the set of highest average feature ranking values arranged by repetitions.

15.2.7 Indicator species analysis for feature selection

For comparison with outcomes of feature selection identified by the previously elaborated machine learning methods, indicator species analysis is employed.[24]

Indicator species (for accuracy, 'taxa' replace 'species') are characterized as key features revealing detrended correspondence analysis (DCA) changes through different samples. A correlational value is evaluated for every feature, considering it to be the outcome of fidelity and specificity for a certain taxon to low or high DCA samples.[25] The extent to which a taxon is linked with an individual marker (for example, low or high DCA) is measured by specificity, and the frequency of a taxon's association with that marker is measured by fidelity. If a taxon only persisted in sites with a certain marker, specificity would have been increased to the maximum, and fidelity would have been increased in the presence of a taxon on each site linked to a certain marker. Every feature is attributed a confidence score employing the bootstrap algorithm, which contrasts the correlational value for every feature selected using the right markers with correlations selected using arbitrarily attributed markers. A low confidence score for the feature–site correlation is caused when the correlation statistics between the site markers and features selected using arbitrary markers are not persistently low in comparison to the correlation statistic employing the right markers. Only taxa with a minimum of 95% confidence (characteristics with relative values higher than 95% of relations distinguished by arbitrary markers) are considered in certain studies. Using the techniques mentioned in Dufrene and Legendre,[25] indicator taxa analysis can be conducted in Python 3.7.[9,25]

15.3 CONCLUSION

An enhanced focus on the research regarding microbial ecology of cold ecosystems is noticed due to high viscosity and low thermal energy with decreased metabolic flux. Microbial groups are in abundance and frequently affect the macroscopic qualities of the ecosystem in which they live. Nonetheless, interpreting the active relationship between certain microbes and ecosystem qualities is a constant trial due to the intricateness of groups. This problem can be tackled by combining advancements in DNA sequence technology with computational methods like machine learning. Applied machine learning incorporates assessing and developing models that employ algorithms having the capacity of identifying, organizing and predicting certain results from data based on microbial ecology. Briefly, the progression of machine learning methods is making a revolution in the area of microbial ecology. Many eminent concerns should be kept in mind while applying machine learning methods to marker-gene datasets, notwithstanding the potential of machine learning algorithms and exceptional sophistication.

REFERENCES

1. Jansson, Janet K., and Kirsten S. Hofmockel. "Soil microbiomes and climate change." *Nature Reviews Microbiology* 18, no. 1 (2020): 35–46.
2. Rizvi, Asfa, Bilal Ahmed, Mohammad Saghir Khan et al. "Psychrophilic bacterial phosphate-biofertilizers: A novel extremophile for sustainable crop production under cold environment." *Microorganisms* 9, no. 12 (2021): 2451.

3. D'Amico, Salvino, Tony Collins, Jean-Claude Marx, Georges Feller, Charles Gerday, and Charles Gerday. "Psychrophilic microorganisms: Challenges for life." *EMBO Reports* 7, no. 4 (2006): 385–389.
4. Morita, Richard Y. "Psychrophilic bacteria." *Bacteriological Reviews* 39, no. 2 (1975): 144–167.
5. Xu, Ying, Yuichi Nogi, Chiaki Kato et al. "*Moritella profunda* sp. nov. and *Moritella abyssi* sp. nov., two psychropiezophilic organisms isolated from deep Atlantic sediments." *International Journal of Systematic and Evolutionary Microbiology* 53, no. 2 (2003): 533–538.
6. Yu, Lei, and Huan Liu. "Feature selection for high-dimensional data: A fast correlation-based filter solution." In Proceedings of the 20th International Conference on Machine Learning (ICML-03), pp. 856–863. 2003.
7. Abramson, N., D. Braverman, and G. Sebestyen. "Pattern recognition and machine learning vol. 9." (1963), Springer.
8. Statnikov, Alexander, Mikael Henaff, Varun Narendra et al. "A comprehensive evaluation of multicategory classification methods for microbiomic data." *Microbiome* 1, no. 1 (2013): 1–12.
9. Thompson, Jaron, Renee Johansen, John Dunbar, and Brian Munsky. "Machine learning to predict microbial community functions: An analysis of dissolved organic carbon from litter decomposition." *PLoS One* 14, no. 7 (2019): e0215502.
10. Sathya, Ramadass, and Annamma Abraham. "Comparison of supervised and unsupervised learning algorithms for pattern classification." *International Journal of Advanced Research in Artificial Intelligence* 2, no. 2 (2013): 34–38.
11. Ruppert, David. "The elements of statistical learning: Data mining, inference, and prediction." (2004): 567–567, Springer.
12. Webb, Hattie E., Dayna M. Brichta-Harhay, Mindy M. Brashears et al. "*Salmonella* in peripheral lymph nodes of healthy cattle at slaughter." *Frontiers in Microbiology* 8 (2017): 2214.
13. Sakowski, Eric, Gherman Uritskiy, Rachel Cooper et al. "Current state of and future opportunities for prediction in microbiome research: Report from the Mid-Atlantic microbiome meet-up in Baltimore on 9 January 2019." *Msystems* 4, no. 5 (2019): e00392-19.
14. Ching, Travers, Daniel S. Himmelstein, Brett K. Beaulieu-Jones et al. "Opportunities and obstacles for deep learning in biology and medicine." *Journal of the Royal Society Interface* 15 (2018): 2017038.
15. Li, Yu, Chao Huang, Lizhong Ding et al. "Deep learning in bioinformatics: Introduction, application, and perspective in the big data era." *Methods* 166 (2019): 4–21.
16. Xu, Hongteng, Dixin Luo, Ricardo Henao et al. "Learning autoencoders with relational regularization." In International Conference on Machine Learning, pp. 10576–10586. PMLR, 2020.
17. LaPierre, Nathan, Chelsea J-T. Ju, G. Zhou et al. "MetaPheno: A critical evaluation of deep learning and machine learning in metagenome-based disease prediction." *Methods* 166 (2019): 74–82.
18. Hunter, Mitchell C., Richard G. Smith, Meagan E. Schipanski et al. "Agriculture in 2050: Recalibrating targets for sustainable intensification." *Bioscience* 67, no. 4 (2017): 386–391.
19. Prakash, Kolla Bhanu, and G. R. Kanagachidambaresan. "Pattern recognition and machine learning." In *Programming with TensorFlow*, pp. 105–144. Springer, Cham, 2021.

20. Team, The Theano Development, Rami Al-Rfou, Guillaume Alain et al. "Theano: A Python framework for fast computation of mathematical expressions." *arXiv preprint arXiv:1605.02688* (2016).
21. Li, Yifeng, Chih-Yu Chen, and Wyeth W. Wasserman. "Deep feature selection: Theory and application to identify enhancers and promoters." *Journal of Computational Biology* 23, no. 5 (2016): 322–336.
22. Pedregosa, Fabian, Gaël Varoquaux, Alexandre Gramfort et al. "Scikit-learn: Machine learning in Python." *The Journal of Machine Learning Research* 12 (2011): 2825–2830.
23. Breiman, Leo, Jerome H. Friedman, Richard A. Olshen, and Charles J. Stone. *Classification and Regression Trees*, pp. 1–368. Edition 1. Routledge, London, 2017.
24. Cáceres, Miquel De, and Pierre Legendre. "Associations between species and groups of sites: Indices and statistical inference." *Ecology* 90, no. 12 (2009): 3566–3574.
25. Dufrêne, Marc, and Pierre Legendre. "Species assemblages and indicator species: The need for a flexible asymmetrical approach." *Ecological Monographs* 67, no. 3 (1997): 345–366.

16 Mass spectrometry as an emerging analytical technique for agriculture under changing climate

Sahar Mumtaz, Fatima Batool, Saliha Khalid, Abida Parveen, Amara Hassan, Mobeen, Muhmmad Arif, Amanullah, Muhammad Adnan, Muhammad Zamin, Abdel Rahman Al Tawaha, Shah Fahad, Fazli Wahid, Amjad Ali, Iqbal Munir and Muhammad Hamzha Saleem**

CONTENTS

* Corresponding author's e-mails: sahar_botany@yahoo.com; saleemhamza312@webmail.hzau.edu.cn

DOI: 10.1201/9781003354031-16

16.1 INTRODUCTION

Analytical chromatography has long been regarded as a potent and effective technique for analysing tiny compounds in complicated matrices in a variety of applications. Gas chromatography (GC) has been described as the best choice among chromatographic methods for the characterisation of volatile and semi-volatile small compounds in diverse matrices/solutions.[1] Under stress-based ecosystems, sustainable agriculture faces an issue of not having easy access to fertiliser and water. It will take good management to acquire enough food for the ever-increasing population from the existing land and water resources, which will necessitate the use of modern advances in agricultural science.[2] To fulfil the growing needs for food, energy and water, agriculture and associated industries will rely on cutting-edge molecular and microscopic instruments. Biosolids, wastewater effluents, manures and fertilisers are among the external inputs used to improve crop productivity. As a result, enhanced analytical instruments such as high-resolution mass spectrometry (HRMS) with GC are needed to determine the mechanism of transport, absorption and accumulation of nutrients and toxicants in the soil–water–plant continuum.[2] It is necessary to become acquainted with and adapted to the evolving situation of scientific instruments and technology in ecosystems. The agricultural and related sectors' cutting-edge technology expands their fecundity to satisfy food, energy and water demands. The protection of natural resources and their management with the least amount of external input has become a problem for the scientific community. To solve the problem, advanced analytical instruments such as HRMS with GC and gas chromatography-tandem mass spectrometry (GC/MS/MS) must be used to understand the realistic mechanism of transport, absorption and accumulation of nutrients and toxicants in the soil–water–plant continuum.[3] Colloidal contaminants include engineered nanoparticles, pathogenic microorganisms (viruses, bacteria and protozoa), colloid-associated phosphorus, radionuclides, heavy metals and organic toxins, all of which can be analysed using advanced and precise analytical tools like single particle inductively coupled plasma-mass spectrometry.[4,5] Furthermore, in the field of forensic toxicology, enantiomeric separation by GC was usually done using indirect methods for medicines and illegal substances, whereas direct methods were preferred for pesticides and polychlorinated biphenyl compounds. Food analysis has mostly used direct methods for chiral separation, whereas environmental analysis has used both direct and indirect approaches, depending on the type of substances examined.[6]

16.2 TECHNIQUES FOR ANALYSING SOIL HEALTH

Soil is a very complex and diverse open three-phase physical medium that is dynamic and in constant change, impacted by environmental, man-made and endogenous influences.[7] Agricultural intensification is putting enormous strain on the soil's ability to retain its functions, resulting in long-term ecosystem deterioration and productivity loss. Hence, there is a pressing need to identify early warning signs. As a result of agricultural management, soil health has deteriorated.[8,9] Soil health refers to a soil's ability to function within natural or controlled limits of the ecosystem to support plant production, water and air quality, human well-being and biodiversity habitats.[10] The need to fulfil the food, fibre and fuel demands of an ever-increasing population has a significant influence on soil health. However, there is growing worry that agricultural intensification is putting enormous strain on the soil's ability to sustain its other functions, potentially leading to large-scale ecosystem deterioration and productivity loss in the long run.[11] Markers of chemical, physical and biological (including microbial) components must be considered in order to assess soil health. Cultivation is recognised to diminish the quantity of organic matter in the soil, therefore reducing nutritional availability.[12] Molecular knowledge of linked biological and geochemical processes is essential for measuring the state of the soil. As a result, a variety of molecular and cellular approaches are now being employed in tandem with biological and chemical markers to improve the ability to assess soil health. These techniques involve various techniques like fluorescence microscopy, *in situ* hybridisation, isotope probing, microbial resilience, DNA quantification denaturing gradient gel electrophoresis, and spectroscopic techniques like MS and GC.[13]

16.3 GAS CHROMATOGRAPHY/MASS SPECTROMETRY

GC/MS is a very important tool used in chemistry to split combinations of chemicals and also categorise the components at their molecular level.[14] Ecological or environmental samples are analysed very accurately by this tool. GC/MS is one of the most used methods for the assessment of volatile analytes. Well-developed gas chromatographic approaches are utilised to analyse volatile analytes.[15] Advances in comprehensive GC enable us to identify and quantify elements at a very basic level. In MS, co-elution could be possible if the detector selectively differentiates the analyte components in the dimension of flow.[16]

GC/MS devices are of different types. Quadrupole-based MS is one of the most common mass spectrometry devices.[17] More complicated integration and multidimensional splitting in MS involve more complicated devices, e.g. GC-TOF-MS (where TOF stands for time of flight). Employment of high-resolution orbital

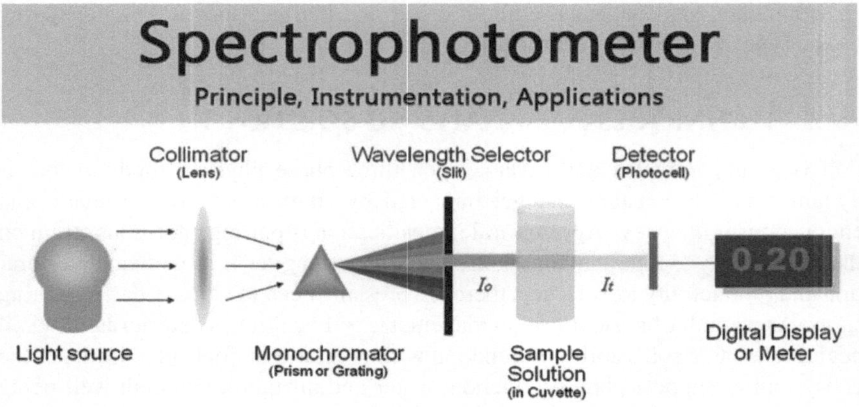

FIGURE 16.1 Working principle of a spectrophotometer.

trapping techniques is the most advanced development in recent GC/MS systems (GC Orbitrap instruments) as shown in Figure 16.1. This most advanced system enables the scientists to detect and identify metabolites and molecules.[18]

Sample preparation is a very complicated task in this advanced technique, which results in the limited suitability of molecules for this analysis. GC/MS takes a very long time for separation of components, which results in reduced utilisation of this technique. But, GC/MS is a very useful device in molecular profiling and metabolic phenotyping studies, and scientists follow one of the two categories of this analysis.[19] Gas chromatographic specialists pay full attention to the optimisation of the chromatographic process and ensure the utilisation of this potential tool. But, this tool can be valuable if the ion source obtains totally separated solutes, qualified by mass spectrometry libraries.[20]

16.4 RECENT APPLICATIONS OF GC/MS

GC/MS is a very useful instrument in analytical chemistry, which separates complex mixtures into their basic elements at the molecular level. Environmental samples can be explored with the highest precision by using this technique. GC operates on the principle that a chemical mixture splits into components when the proper heat is applied.[21] A column is utilised to carry the heated gases along with an inert gas (helium). This process results in the separation of components, which arrive at the column opening and then drift into the mass spectrometer.[22] Identification of individual components is carried out by the mass of elements utilising MS. A library of mass spectra of numerous compounds is established on a computer.

16.4.1 DNA QUANTIFICATION

GC/MS is very beneficial for the quantification of oxidised DNA because it oxidises all the base pairs of DNA. Derivation of DNA base pairs is necessary prior to the quantification by GC/MS.[23] It is advised by many scientists that DNA nucleosides must be kept separated from oxidised nucleosides when utilising column chromatography. It must be certain that the derivatisation process includes only oxidised base pairs. Such advances in GC/MS have resulted in the development of innovative approaches such as high-performance liquid chromatography (HPLC).[24]

16.4.2 Metabolite measurement

GC/MS is a well-recognised analytical tool used widely for both biological and environmental studies.[25] This analytical tool is perfectly suited for the splitting of volatile chemicals. GC/MS is a very useful technique to separate the polar species that occur in human metabolites. Most of the building blocks of macromolecules, like amino acids and fatty acids, found in serum and urine are identified by the employment of GC/MS. This device occupies a vital position in the diagnosis of several metabolic and inherited diseases.[26]

16.4.3 Alteration of organic matter

The carbon cycle of terrestrial ecosystems is modified by vast fires of vegetation through the development of stable igneous structures. GC/MS is utilised to study fire-induced alterations in the structure of organic matter.[27] This technique was employed to characterise the chemistry of organic matter in the litter layer of a wood that had gone through a fire. Thermo-sensitive paint was used to monitor the temperature peaks of the fire. Samples were subjected to size fractionation before and after the fire based on burnt and unburnt proportions, such as black and brown. Carbon and nitrogen contents as well as stable carbon isotopes were also analysed by GC/MS.[28] Solid-state ^{13}C nuclear magnetic resonance spectroscopy was used to evaluate the components of >2 mm in size. All these comparisons were made between black and brown post-fire litter in >370 °C plots.

16.4.4 Determination of ergosterol as an indicator of fungal biomass

A fungal disease named fusarium head blight is triggered by some species of genus *Fusarium* that infect all the varieties of wheat and barley in the USA. Reduced yield and poor-quality grains are some of the consequences of this disease.[29] These fungal diseases can contaminate the grains produced by these plants, resulting in decreased food for humans and animals. The fungal cell membrane is mainly made up of ergosterol. This can also be present in higher plants and is widely used as a chemical marker to determine fungal presence in plant tissues, grains, airborne dust and soil. The presence of ergosterol in building material and airborne dust is determined by GC/MS due to the higher level of detection accuracy.[30] The detection of ergosterol in grains was conducted by HPLC. The best method to quantify ergosterol in wheat and barley is through GC/MS.

16.4.5 Characterisation and evaluation of smoke tracers in particulate matter

Particulate matter of smoke is a combination of a variety of particles resulting from combustion. Wood smoke results in the abundant production of methoxyphenols, and their presence in air is distinctive to biomass combustion.[31,32] Methoxyphenols are emitted from the wood in both particles and vapours. The combustion of biomass is the main source of particulate matter present in the atmosphere. Particulate matter produced due to biomass combustion is mostly of organic nature. Samples of particulate matter are collected after wild land fires for analytical comparison of different particles using quartz filters. Liquid-based separation methods and GC/MS are employed to determine the comparative concentration of different particles.[33] The GC/MS technique is more accurate regarding selectivity and sensitivity as compared with liquid-based separation methods. Some laboratories have employed HPLC to analyse smoke tracers of particulate matter in comparison with GC/MS.

16.4.6 Conversion of rice husks and sawdust to liquid fuel via pyrolysis

Pyrolysis is a process in which biomass is thermally converted to high-energy materials like charcoal, oil and gas at a temperature lower than 600 °C with no oxygen.[34] The production of gases mostly occurs above 600 °C along with the presence of oxygen. Gases formed in this process are methane, hydrogen, carbon dioxide, carbon monoxide and nitrogen. Thermal conversion of biomass into liquid fuel is one of the oldest methods in bioenergy production. Pyrolysis is the main reaction of thermal conversion because all the reactions starting at 150–200 °C are irreversible in the absence of oxygen.[35] At a lower temperature of pyrolysis, oxygen cannot enter this reaction, because the gases produced in this process surround the reaction and protect it from oxygen. The products (gas, liquid fuel, coal) produced as the result of pyrolysis depend upon the conditions and the availability of oxygen.

16.4.7 Coal pyrolysis and hydropyrolysis

The products of coal pyrolysis have been investigated at the molecular level to enhance the non-fuel use of coal. Some of the depositional factors of coal affect the molecular composition of coal pyrolysis.[36] It's not possible to investigate the composition of coal pyrolysis due to the highly complex structure of coal. Pyrolysis-GC/MS is a very advanced technique utilised to investigate the molecular composition of coal pyrolysis products. Hierarchical cluster analysis, chemometric tools and principal component analysis (PCA) showed correlation among products of coal pyrolysis. The results of the molecular composition of coal pyrolysis can be employed for the evaluation of coal tar from hydropyrolysis.[37] Coal can be very useful in industry when its molecular composition is interpreted by employing pyrolysis-GC/MS along with the application of chemometrics.

16.4.8 Soot formation

Combustion is a process in which soot is produced along with energy, which is very problematic for scientists and engineers. Soot production is an indication of incomplete combustion and low efficiency of the combustion process.[38] It may result in reduced visibility and enhanced particulate matter in the atmosphere. Heavy soot and smoke emission cause enhanced particulate loading in our environment, which imparts unpleasant effects on human health. Pyrolysis-GC/MS is employed to detect and quantify particulate matter in soot.[39] Soot or smoke particulates have a deleterious effect on the efficiency of devices, so in advanced practices, designers avoid smoke formation. In advanced designing, the technological problem is to produce smoke in such a way that particles can be oxidised before they leave the furnace.

16.4.9 Desorption of surface oxides up to 1100 °C

Oxides are catalytically active compounds that are most commonly utilised as support material. These components have great technical importance, but very little work has been done in systematic surface science.[40] Well-documented data is present regarding adsorption of oxide powders, but microscopic models of these procedures are still missing. To understand the process of surface oxide desorption, microscopic models are required. For this purpose, HPLC and GC/MS were employed to detect the desorption of surface oxides at 1100 °C. These techniques provide understanding about electronic structures of surface oxides and their interaction with the substrate. The surface science method has one problem: it hinders the utilisation of electron spectroscopic methods. A single, thin oxide film can be developed on a metallic surface to overcome this problem.

16.4.10 Temperature-programmed desorption of young chars at 1650 °C

Deactivation of young chars can be achieved when they are kept in higher-temperature conditions. The amount of surface oxides on young chars was examined at 1700 °C by temperature-programmed desorption.[41] Surface oxides are located abundantly on young chars as compared with old chars at higher temperature. Chars that are oxidised at 629 °C exhibit different products at three different temperature ranges. Peaks of temperature-programmed desorption at 725 °C result in an energy ranged of 170 kJ mol^{-1}. activated At a temperature of 1430 °C, C 300 kJ mol^{-1} energy was activated when carbon mono-oxide desorbed. Aluminium tubes were converted to Al_2O and a considerable amount of oxygen emitted at 1300 °C.[42] Furthermore, aluminium tubes and related material form

CO_2 by reacting with CO and are also able to form CO by reacting with carbon. GC/MS was used to detect desorption of oxides of young chars at 1650 °C.

16.4.11 Isotope labelling techniques

For the quantification of metabolic activities, stable isotope tracers are utilised. The 2H-water labelling technique is very useful because it incorporates several macromolecules that enable the measurement of lipids, carbohydrates, nucleic acids and proteins.[43] There is a limitation to 2H-water administration because the measurement of small changes in isotopic enrichment is challenging. GC/MS determines the isotopic ratio and benefits of isotope fractionation. The degree of mass spectrometer, chromatographic peaks and isotopic fractionation can enhance isotope labelling.[44,45] GC/MS can measure asymmetrical peaks of 2H-water labelling at a low signal background.

16.5 CONCLUSION AND FUTURE OUTLOOK

To fulfil the growing needs for food, energy and water, agriculture and associated industries rely on molecular and microscopic instruments. Food and agriculture laboratories regularly use various chromatographic methods to analyse the safety and purity of goods. From the invention of GC/MS to today's advances, the quality of this technique has been improved. Currently, the topics of focus are agriculture, biochemistry, pharmacology, metabolomics and medical analysis. HRMS, softer ionisation techniques and multidimensionality are some of the endeavours that still need to be achieved in GC/MS. Furthermore, the application of a standardised workflow will allow the utilisation of the full potential of GC/MS in the future.

ACKNOWLEDGEMENT

We are thankful to the University of Education Lahore for provision of facilities in writing this chapter.

REFERENCES

1. Fitz, Brian D., Ryan B. Wilson, Brendon A. Parsons et al. "Fast, high peak capacity separations in comprehensive two-dimensional gas chromatography with time-of-flight mass spectrometry." *Journal of Chromatography* 1266 (2012): 116–123.
2. Richardson, Susan D., and Thomas A. Ternes. "Water analysis: Emerging contaminants and current issues." *Analytical Chemistry* 86, no. 6 (2014): 2813–2848.
3. Raghavendra, M., M. P. Sharma, A. Ramesh et al. "Soil health indicators: Methods and applications." In *Soil Analysis: Recent Trends and Applications*, pp. 221–253. Springer, Singapore, 2020.
4. Mitrano, Denise M., Angela Barber, Anthony Bednar et al. "Silver nanoparticle characterization using single particle ICP-MS (SP-ICP-MS) and asymmetrical flow field flow fractionation ICP-MS (AF4-ICP-MS)." *Journal of Analytical Atomic Spectrometry* 27, no. 7 (2012): 1131–1142.

5. Adnan, Muhammad, Zahir Shah, Muhammad Sharif, and Hidayatur Rahman. "Liming induces carbon dioxide (CO2) emission in PSB inoculated alkaline soil supplemented with different phosphorus sources." *Environmental Science and Pollution Research* 25, no. 10 (2018): 9501–9509.
6. Ribeiro, Cláudia, Ricardo Gonçalves, and M. E. Tiritan. "Separation of enantiomers using gas chromatography: Application in forensic toxicology, food and environmental analysis." *Critical Reviews in Analytical Chemistry* 51, no. 8 (2021): 787–811.
7. Rodríguez-Añón, J., J. Proupín-Castiñeiras, M. Villanueva-López, and O. Núñez-Fernández. "Development of an experimental procedure to analyse the 'soil health state' by microcalorimetry." *Journal of Thermal Analysis and Calorimetry* 87, no. 1 (2007): 15–19.
8. Khan, Ahsan, Ali Muhammad Yousafzai, Nazish Shah et al. "Enzymatic profile activity of grass carp (Ctenopharyngodon idella) after exposure to the pollutant named atrazine (herbicide)." *Polish Journal of Environmental Studies* 25, no. 5 (2016).
9. Odhong', Charles, Andreas Wilkes, Suzanne van Dijk et al. "Financing large-scale mitigation by smallholder farmers: what roles for public climate finance?." *Frontiers in Sustainable Food Systems* 3 (2019): 3.
10. Gugino, B. K., O. J. Idowu, R. R. Schindelbeck et al. *Cornell Soil Health Assessment Training Manual*, Edition 2.0. Cornell University, Geneva, 2009.
11. Vitousek, Peter M., Rosamond Naylor, Timothy Crews et al. "Nutrient imbalances in agricultural development." *Science* 324, no. 5934 (2009): 1519–1520.
12. Schlesinger, William H., and Emily S. Bernhardt. *Biogeochemistry: An Analysis of Global Change*. Academic Press, 2013.
13. Arias, M. Enriqueta, José A. González-Pérez, Francisco J. González-Vila, and Andrew S. Ball. "Soil health: A new challenge for microbiologists and chemists." *International Microbiology* 8, no. 1 (2005): 13–21.
14. Bowman, David T., Karl J. Jobst, Xavier Ortiz et al. "Improved coverage of naphthenic acid fraction compounds by comprehensive two-dimensional gas chromatography coupled with high resolution mass spectrometry." *Journal of Chromatography A* 1536 (2018): 88–95.
15. Fiehn, Oliver. "Metabolomics by gas chromatography–mass spectrometry: Combined targeted and untargeted profiling." *Current Protocols in Molecular Biology* 114, no. 1 (2016): 30–34.
16. Franchina, Flavio Antonio, Delphine Zanella, Eliane Lazzari et al. "Investigating aroma diversity combining purge-and-trap, comprehensive two-dimensional gas chromatography, and mass spectrometry." *Journal of Separation Science* 43, no. 9–10 (2020): 1790–1799.
17. Hayward, Douglas G., Jeffery C. Archer, Sue Andrews et al. "Application of a high-resolution quadrupole/orbital trapping mass spectrometer coupled to a gas chromatograph for the determination of persistent organic pollutants in cow's and human milk." *Journal of Agricultural and Food Chemistry* 66, no. 44 (2018): 11823–11829.
18. García-Bermejo, Ángel, Manuela Ábalos, Jordi Sauló et al. "Triple quadrupole tandem mass spectrometry: A real alternative to high resolution magnetic sector instrument for the analysis of polychlorinated dibenzo-p-dioxins, furans and dioxin-like polychlorinated biphenyls." *Analytica Chimica Acta* 889 (2015): 156–165.
19. Santini, Giuseppe, Nadia Mores, Andreu Penas et al. "Electronic nose and exhaled breath NMR-based metabolomics applications in airways disease." *Current Topics in Medicinal Chemistry* 16, no. 14 (2016): 1610–1630.

20. Hu, Wenjing, Hui Zhang, and Bingjian Zhang. "Identification of organic binders in ancient Chinese paintings by immunological techniques." *Microscopy and Microanalysis* 21, no. 5 (2015): 1278–1287.
21. Pierce, Karisa M., Janiece L. Hope, Jamin C. Hoggard, and Robert E. Synovec. "A principal component analysis based method to discover chemical differences in comprehensive two-dimensional gas chromatography with time-of-flight mass spectrometry (GC× GC-TOFMS) separations of metabolites in plant samples." *Talanta* 70, no. 4 (2006): 797–804.
22. Tranchida, Peter Q., Giorgia Purcaro, Mariarosa Maimone, and Luigi Mondello. "Impact of comprehensive two-dimensional gas chromatography with mass spectrometry on food analysis." *Journal of Separation Science* 39, no. 1 (2016): 149–161.
23. Dizdaroglu, Miral, Pawel Jaruga, Mustafa Birincioglu, and Henry Rodriguez. "Free radical-induced damage to DNA: Mechanisms and measurement." *Free Radical Biology and Medicine* 32, no. 11 (2002): 1102–1115.
24. Aybastıer, Önder, Sam Dawbaa, Cevdet Demir et al. "Quantification of DNA damage products by gas chromatography tandem mass spectrometry in lung cell lines and prevention effect of thyme antioxidants on oxidative induced DNA damage." *Mutation Research/Fundamental and Molecular Mechanisms of Mutagenesis* 808 (2018): 1–9.
25. Díaz-Ruiz-Navamuel, Emma, Andrés Ortega Piris, and Carlos A. Pérez-Labajos. "Reduction in CO2 emissions in RoRo/Pax ports equipped with automatic mooring systems." *Environmental Pollution* 241 (2018): 879–886.
26. Sato, Hirotaka, Yuki Ito, Miho Inoue et al. "Simple method to detect triclofos and its metabolites in plasma of children by combined use of liquid chromatography tandem-mass spectrometry and gas chromatography-mass spectrometry." *Scientific Reports* 9, no. 1 (2019): 1–9.
27. Watanabe, Ayano, Keiki Okazaki, Toshihiro Watanabe et al. "Metabolite profiling of mizuna (*Brassica rapa* L. var. Nipponsinica) to evaluate the effects of organic matter amendments." *Journal of Agricultural and Food Chemistry* 61, no. 5 (2013): 1009–1016.
28. Jeanneau, Laurent, Emilie Jardé, and Gérard Gruau. "Influence of salinity and natural organic matter on the solid phase extraction of sterols and stanols: Application to the determination of the human sterol fingerprint in aqueous matrices." *Journal of Chromatography A* 1218, no. 18 (2011): 2513–2520.
29. Parsi, Ziba, and Tadeusz Górecki. "Determination of ergosterol as an indicator of fungal biomass in various samples using non-discriminating flash pyrolysis." *Journal of Chromatography A* 1130, no. 1 (2006): 145–150.
30. Mille-Lindblom, Cecilia, Eddie Von Wachenfeldt, and Lars J. Tranvik. "Ergosterol as a measure of living fungal biomass: Persistence in environmental samples after fungal death." *Journal of Microbiological Methods* 59, no. 2 (2004): 253–262.
31. Uddin, Muhammad Nazir, Murtaza Ali, Muhammad Farooq Muhammad et al. "Characterizing microbial populations in petroleum-contaminated soils of Swat District, Pakistan." *Pol J Env Stud* 25, no. 4 (2016): 1721–1727.
32. Verma, Vishal, Andrea Polidori, James J. Schauer et al. "Physicochemical and toxicological profiles of particulate matter in Los Angeles during the October 2007 southern California wildfires." *Environmental Science & Technology* 43, no. 3 (2009): 954–960.
33. Xu, Hongmei, Benjamin Guinot, Steven Sai Hang Ho et al. "Evaluation on exposures to particulate matter at a junior secondary school: A comprehensive study on health risks and effective inflammatory responses in Northwestern China." *Environmental Geochemistry and Health* 40, no. 2 (2018): 849–863.

34. Zheng, Ji-lu, Xi-feng Zhu, Qing-xiang Guo, and Qing-shi Zhu. "Thermal conversion of rice husks and sawdust to liquid fuel." *Waste Management* 26, no. 12 (2006): 1430–1435.
35. Kumagai, Seiji, Yosuke Noguchi, Yasuji Kurimoto, and Koichi Takeda. "Oil adsorbent produced by the carbonization of rice husks." *Waste Management* 27, no. 4 (2007): 554–561.
36. Zhang, Jie, Nan Zheng, and Jie Wang. "Comparative investigation of rice husk, thermoplastic bituminous coal and their blends in production of value-added gaseous and liquid products during hydropyrolysis/co-hydropyrolysis." *Bioresource Technology* 268 (2018): 445–453.
37. Zheng, Nan, Jie Zhang, and Jie Wang. "Parametric study of two-stage hydropyrolysis of lignocellulosic biomass for production of gaseous and light aromatic hydrocarbons." *Bioresource Technology* 244 (2017): 142–150.
38. Wang, Xuebin, Qiming Jin, and Liang Wang, "Synergistic effect of biomass and polyurethane waste co-pyrolysis on soot formation at high temperatures." *Journal of Environmental Management* 239 (2019): 306–315.
39. Baldelli, Alberto, Una Trivanovic, Timothy A. Sipkens, and Steven N. Rogak. "On determining soot maturity: A review of the role of microscopy-and spectroscopy-based techniques." *Chemosphere* 252 (2020): 126532.
40. You, Yajie, Yuzhen Liang, Shimeng Peng et al. "Modeling coupled kinetics of arsenic adsorption/desorption and oxidation in ferrihydrite-Mn (II)/manganese (oxyhydr) oxides systems." *Chemosphere* 244 (2020): 125517.
41. Rüfer, Corinna E., Jutta Moeseneder, Karlis Briviba et al. "Bioavailability of astaxanthin stereoisomers from wild (Oncorhynchus spp.) and aquacultured (*Salmo salar*) salmon in healthy men: A randomised, double-blind study." *British Journal of Nutrition* 99, no. 5 (2008): 1048–1054.
42. Liu, Qing, Niladri Basu, Giles Goetz et al. "Differential gene expression associated with dietary methylmercury (MeHg) exposure in rainbow trout (*Oncorhynchus mykiss*) and zebrafish (Danio rerio)." *Ecotoxicology* 22, no. 4 (2013): 740–751.
43. Obermaier, Christian, Anja Griebel, and Reiner Westermeier. "Principles of protein labeling techniques." In *Proteomic Profiling*, pp. 153–165. Humana Press, New York, 2015.
44. Asadi-Atoi, Paria, Pierre Barraud, Carine Tisne, and Stefanie Kellner. "Benefits of stable isotope labeling in RNA analysis." *Biological Chemistry* 400, no. 7 (2019): 847–865.
45. Gruber, Beate, Frank David, and Pat Sandra. "Capillary gas chromatography-mass spectrometry: Current trends and perspectives." *Trends in Analytical Chemistry* 124 (2020): 115475.

Index

C

D

N

O

P

T

U

W